Quantum Metaphysical Investigations : Exploring the Depths of Reality and Consciousness

Marino Baca-Carmona

Table of Contents

Chapter 1: Foundations of quantum metaphysics

- Questions 1-10: Exploring the quantum nature of reality

- Answers 1-10: Unraveling the mysteries of quantum existence

Chapter 2: Advanced Concepts in Quantum Cosmology

- Questions 11-20: Delving into the quantum origins of the universe

- Answers 11-20: Navigating the complexities of quantum cosmological theories

Chapter 3: Quantum Epistemology and Cognition

- Questions 21-30: Probing the interface between quantum phenomena and consciousness

- Answers 21-30: Unlocking the secrets of quantum cognition and epistemology

Chapter 4: Quantum ontological emergence

- Questions 31-40: Investigating the emergence of new ontological structures

Quantum Metaphysical Investigations : Exploring the Depths of Reality and Consciousness
Marino Baca-Carmona

- Answers 31-40: Tracing the paths of quantum ontological emergence

Chapter 5: Metaphysical implications of quantum theory

- Questions 41-50: Examination of the metaphysical ramifications of quantum mechanics

- Answers 41-50: Exposing the metaphysical importance of quantum theory

Chapter 6: Quantum Consciousness and Mind

- Questions 51-60: Exploring the quantum nature of consciousness

- Answers 51-60: Shedding light on the enigma of quantum consciousness

Chapter 7: Quantum cosmogony and multiverse

- Questions 61-70: Contemplating quantum origins and the nature of the multiverse

- Answers 61-70: Unraveling the mysteries of quantum cosmogony and the multiverse

Chapter 8: Transcendent Realities and Hypercognition

Quantum Metaphysical Investigations : Exploring the Depths of Reality and Consciousness

Marino Baca-Carmona

- Questions 71-80: Venturing into transcendent realms and hypercognitive domains

- Answers 71-80: Illuminating the boundaries of transcendent realities and hypercognition

Chapter 9: Metaphysical Inquiries and Quantum Epistemology

- Questions 81-90: Engage in meta-metaphysical speculations and quantum epistemological investigations

- Answers 81-90: Delve into the depths of metametaphysical investigations and quantum epistemology

Chapter 10: Quantum ontological complexity and hyperreality

- Questions 91-100: Fight against quantum ontological complexity and hyperrealities

- Answers 91-100: Revealing the complexity of quantum ontological realities and hyperrealities

Chapter 11: Quantum Hypermetaphysical Transcendence

- Questions 101-110: Exploring the limits of the quantum hypermetaphysical realms

Quantum Metaphysical Investigations : Exploring the Depths of Reality and Consciousness
Marino Baca-Carmona

- Answers 101-110: Delve into quantum hypermetaphysical transcendence

Chapter 12: Hypercognitive Quantum Cosmology

- Questions 111-120: Investigation of hypercognitive cosmological theories

- Answers 111-120: Illuminating the complexities of hypercognitive quantum cosmology

Chapter 13: Quantum Trans-Meta-Epistemological Perspectives

- Questions 121-130: Delving into quantum transmetaepistemological queries

- Answers 121-130: Exposition of quantum transmeta-epistemological perspectives

Chapter 14: Quantum Hypertransontological Consciousness

- Questions 131-140: Contemplating the nature of quantum hyper-transontological consciousness

- Answers 131-140: Shedding light on quantum hypertransontological consciousness

Chapter 15: Quantum Metacognitive Realms and Transcendence

Quantum Metaphysical Investigations : Exploring the Depths of Reality and Consciousness
Marino Baca-Carmona

- Questions 141-150: Exploring quantum metacognitive realities and transcendence

- Answers 141-150: Unraveling the mysteries of the quantum metacognitive realms

Chapter 16: Hyperquantum meta-metaphysical epistemology

- Questions 151-160: Participation in hyperquantum meta-metaphysical epistemological research

- Answers 151-160: Delving into hyperquantum metametaphysical epistemology

Chapter 17: Quantum Hyperontological Transcendence

- Questions 161-170: Investigation of quantum hyperontological transcendence

- Answers 161-170: Illuminating the depths of quantum hyperontological transcendence

Chapter 18: Transcendental Quantum Phenomenology

- Questions 171-180: Delve into transcendental quantum phenomenological experiences

Quantum Metaphysical Investigations : <u>Exploring the Depths of Reality and Consciousness</u>
Marino Baca-Carmona

- Answers 171-180: Exploring the mysteries of transcendental quantum phenomenology

Chapter 19: Hypermetacognitive Quantum Realities

- Questions 181-190: Contemplating hypermetacognitive quantum realities

- Answers 181-190: Revealing the complexities of quantum hypermetacognitive realities

Chapter 20: Hyperquantum Transepistemological Cosmogony

- Questions 191-200: Exploration of hyperquantum transepistemological cosmogonic theories

- Answers 191-200: Exposition on the complexities of hyperquantum transepistemological cosmogony

Chapter 21: Hypertranscendental Quantum Ontology

- Questions 201-210: Investigation of the hyper-transcendental quantum nature of reality

- Answers 201-210: They shed light on hypertranscendental quantum ontology

Quantum Metaphysical Investigations : Exploring the Depths of Reality and Consciousness

Marino Baca-Carmona

Chapter 22: Hyperquantum Metaphysical Transcendence

- Questions 211-220: Contemplating hyperquantum metaphysical transcendence

- Answers 211-220: Illuminating the depths of hyperquantum metaphysical transcendence

Chapter 23: Quantum Trans-Hyper-Meta-Epistemological Realities

- Questions 221-230: Delving into the trans-hyper-meta-epistemological quantum realms

- Answers 221-230: Exploring the mysteries of quantum trans-hyper-meta-epistemological realities

Chapter 24: Hyperquantum Meta-Metaphysical Research

- Questions 231-240: Participate in hyperquantum meta-metaphysical research

- Answers 231-240: Unraveling the complexities of hyperquantum meta-metaphysical research

Chapter 25: Quantum hypertransontological phenomena

- Questions 241-250: Investigation of quantum hypertransontological phenomena

Quantum Metaphysical Investigations : Exploring the Depths of Reality and Consciousness
Marino Baca-Carmona

- Answers 241-250: Delving into the enigma of quantum hypertransontological phenomena

Chapter 26: Hyperquantum Transcendental Cognition

- Questions 251-260: Contemplation of hyperquantum transcendental cognitive realities

- Answers 251-260: Exploring the depths of hyperquantum transcendental cognition

Chapter 27: Quantum hyper-trans-epistemological realms

- Questions 261-270: Depth into the quantum hypertrans-epistemological dimensions

- Answers 261-270: Illuminating the mysteries of the quantum hyper-trans-epistemological realms

Chapter 28: Hyperquantum Metacognitive Transcendence

- Questions 271-280: Investigation of hyperquantum metacognitive transcendence

- Answers 271-280: Revealing the complexities of hyperquantum metacognitive transcendence

Quantum Metaphysical Investigations : Exploring the Depths of Reality and Consciousness
Marino Baca-Carmona

Chapter 29: Quantum Trans-Hyper-Meta-Ontological Research

•	Questions 281-290: Engage in quantum transhyper-metaontological research

•	Answers 281-290: Exploring the depths of quantum transhyper-metaontological research

Chapter 30: Hyperquantum Transepistemological Reality

•	Questions 291-300: Contemplating hyperquantum transepistemological realities

•	Answers 291-300: Illuminating the enigma of hyperquantum transepistemological reality

1. What fundamental principles govern the emergence of consciousness within the framework of quantum mechanics and information theory?

2. How do we reconcile the apparent duality of existence, the subjective experience of consciousness and the objective reality described by physics, into a unified theory of everything?

3. Can the human mind fully understand the nature of reality, or are we limited by the limitations of our own cognitive architecture?

Quantum Metaphysical Investigations : Exploring the Depths of Reality and Consciousness
Marino Baca-Carmona

4. In a universe governed by probabilistic quantum mechanics, is free will an illusion or a fundamental aspect of human agency?

5. How do we navigate the tension between the inherent subjectivity of human experience and the objective pursuit of scientific truth?

6. Is there a boundary between the physical and metaphysical realms, or are they interconnected in ways that challenge our conventional understanding of reality?

7. What role does uncertainty play in shaping our perception of reality, and can we truly escape the confines of uncertainty in our pursuit of knowledge?

8. How do cultural, social and historical factors influence our understanding of existential issues, and can we achieve a truly objective perspective on reality?

9. Can artificial intelligence ever achieve genuine consciousness and, if so, what are the ethical implications for humanity's relationship with intelligent machines?

10. Is there a definitive purpose or meaning to existence, or are we simply products of chance and circumstance in an indifferent universe?

11. How do we reconcile the apparent conflict between the deterministic laws of physics and the indeterminacy inherent in human experience, particularly in the context of quantum mechanics and chaos theory?

Quantum Metaphysical Investigations : Exploring the Depths of Reality and Consciousness
Marino Baca-Carmona

12. Can we develop a comprehensive framework that takes into account the interconnectedness of all phenomena, from the microscopic to the macroscopic, incorporating both the physical and metaphysical realms?

13. What implications do the principles of emergence and self-organization have for our understanding of consciousness, particularly in complex systems such as the human brain?

14. Is reality a static construct, or is it in a constant state of flux, shaped by the dynamic interplay of perception, cognition, and external stimuli?

15. How do we navigate the tension between individual autonomy and collective responsibility in a world characterized by interconnection and interdependence?

16. Can we conceive of alternative forms of intelligence that transcend our understanding of human-centered cognition, potentially encompassing non-human and even non-biological entities?

17. What role does language play in shaping our perception of reality, and can we escape the limitations imposed by the structure of language itself?

18. How do we explain the existence of consciousness in a universe that appears to be largely devoid of it, and what implications does this have for our understanding of the cosmos?

Quantum Metaphysical Investigations : Exploring the Depths of Reality and Consciousness
Marino Baca-Carmona

19. Is there a fundamental limit to human knowledge, beyond which we cannot venture, or is the search for knowledge an infinite journey with no final destination?

20. Can we ever achieve a truly objective understanding of reality, or are we forever bound by the subjective limitations of our own consciousness?

21. How do the principles of nonlocality and entanglement in quantum mechanics challenge our understanding of space, time, and the nature of interconnectedness at the most fundamental level of reality?

22. Can the concept of "self" be deconstructed into its constituent elements, revealing the illusion of individual identity and the interconnectedness of all beings?

23. How do we reconcile the existence of suffering and injustice in the world with the idea of a benevolent or omnipotent higher power, and what implications does this have for our moral and ethical frameworks?

24. Is there a deeper purpose or meaning to the seemingly random events and happenings that shape our lives, or are we simply observers in a universe governed by chance and chaos?

25. Can we truly transcend the limitations of our human perspective to perceive reality as it truly is, or are we forever confined to the subjective lens through which we interpret the world?

Quantum Metaphysical Investigations : Exploring the Depths of Reality and Consciousness
Marino Baca-Carmona

26. How do we distinguish between genuine knowledge and mere belief, and what criteria can we use to evaluate the validity and reliability of our understanding of the world?

27. Is there a fundamental difference between the concepts of existence and reality, and if so, how do we define and conceptualize each within the context of our own subjective experience?

28. How do the principles of symmetry and asymmetry manifest themselves in the structure of the universe, and what implications do they have for our understanding of order, chaos, and complexity?

29. Can we conceive of alternative modes of consciousness that transcend our current understanding of consciousness, potentially leading to new ideas about the nature of reality and existence?

30. What is the relationship between time and consciousness, and how do subjective experiences of temporality shape our perception of reality and self?

31. How do we reconcile the paradox of infinite regress and final causation within the framework of cosmology and metaphysics, and what implications does this have for our understanding of the origins of the universe?

32. Can we conceive of a multidimensional model of reality that transcends our conventional understanding of space-time, incorporating higher dimensions and alternative realities?

Quantum Metaphysical Investigations : Exploring the Depths of Reality and Consciousness
Marino Baca-Carmona

33. How do we explain the phenomenon of consciousness arising from seemingly inert matter, and what role do emergent properties play in the evolution of complexity within the universe?

34. Is reality a singular, objective construct, or does it manifest as a multiplicity of subjective experiences, each shaped by the unique perspectives of individual observers?

35. Can we develop a unified theory of consciousness that integrates ideas from neuroscience, psychology, philosophy, and spirituality, providing a comprehensive understanding of the nature of consciousness and perception?

36. How do we distinguish between the concepts of determinism and randomness in the context of quantum mechanics, and what implications do they have for our understanding of free will and determinism?

37. Can the principles of symmetry and asymmetry in nature be reconciled with the existence of fundamental asymmetries such as the arrow of time and the imbalance between matter and antimatter?

38. How do the concepts of impermanence and interconnectedness in Buddhist philosophy intersect with modern scientific theories of reality, and what insights can be gained from exploring their convergence?

39. Is there a fundamental limit to the complexity of the universe, beyond which it becomes incomprehensible to finite human minds, or

does the pursuit of knowledge offer unlimited potential for understanding?

40. Can we ever close the gap between the subjective experience of consciousness and the objective reality described by science, or are they fundamentally irreconcilable aspects of existence?

41. How do we reconcile the paradox of quantum entanglement with the macroscopic world, and what implications does this have for our understanding of the nature of reality and causality?

42. Can we develop a comprehensive theory of consciousness that takes into account the phenomenon of qualia, the subjective qualities of sensory experiences, within the framework of materialist neuroscience?

43. What role does the observer play in shaping the results of quantum experiments, and how does this challenge our understanding of objectivity and the nature of reality?

44. Can we conceive of a meta-level of reality beyond the observable universe and, if so, how might it intersect or influence our own existence?

45. How do the principles of emergence and self-organization give rise to complex systems such as ecosystems, economies, and societies, and what insights can they provide about the nature of order and chaos?

Quantum Metaphysical Investigations : Exploring the Depths of Reality and Consciousness
Marino Baca-Carmona

46. Is there a fundamental limit to human knowledge, beyond which certain truths are inherently unknowable, or can we continue to push the limits of understanding indefinitely?

47. How do the concepts of time dilation and spacetime curvature in general relativity intersect with the subjective experience of time, and what implications does this have for our understanding of temporality and causality?

48. Can we develop a coherent framework for understanding the relationship between mind and matter, incorporating ideas from quantum mechanics, neuroscience, and Eastern philosophies such as Advaita Vedanta?

49. How do the principles of symmetry breaking and phase transitions in physics inform our understanding of the emergence of complexity and diversity in the universe, from subatomic particles to galaxies?

50. Can we ever achieve a unified theory of everything that reconciles the seemingly disparate phenomena of the quantum realm and the gravitational realm, providing a complete description of the fundamental nature of reality?

51. How do the principles of Gödel's incompleteness theorem and Tarski's indefinability theorem challenge our notions of mathematical truth and the foundations of logic, and what implications do they have for our understanding of reality?

Quantum Metaphysical Investigations : Exploring the Depths of Reality and Consciousness
Marino Baca-Carmona

52. Can we conceive of a meta-level of reality beyond the constraints of space-time, where concepts such as causality and locality may not apply, and if so, how might this inform our understanding of existence?

53. How do the principles of non-Euclidean geometry and topology illuminate the structure of the universe, and what insights can they provide about the nature of higher-dimensional spaces?

54. Can we develop a rigorous framework for understanding the relationship between consciousness and the physical brain, incorporating insights from fields such as quantum biology, neurophenomenology, and panpsychism?

55. What role does the concept of information play in shaping our understanding of reality, and how might the universe be viewed as a vast information-processing network?

56. How do the principles of complementarity and contextuality in quantum mechanics challenge our classical intuitions about the nature of reality, and what new perspectives do they offer on the relationship between the observer and the observed?

57. Can we conceive of alternative modes of perception and cognition that transcend the limitations of our human senses, potentially leading to new insights into the fabric of reality?

58. How do the principles of self-reference and recursion in mathematics and computing inform our understanding of the nature of consciousness,

Quantum Metaphysical Investigations : Exploring the Depths of Reality and Consciousness
Marino Baca-Carmona

particularly in relation to concepts such as self-awareness and introspection?

59. Can we develop a unified framework for understanding the relationship between mind, matter and consciousness, incorporating ideas from Eastern philosophies, Western science and transpersonal psychology?

60. How do the principles of emergence and complexity theory shed light on the nature of consciousness, and what implications do they have for our understanding of free will, agency, and the nature of self?

61. How do we reconcile the paradox of the observer effect in quantum mechanics with the macroscopic world, and what implications does this have for our understanding of the nature of reality and the role of consciousness in shaping it?

62. Can we develop a comprehensive theory of cosmogenesis that explains the origin and evolution of the universe from a state of primordial chaos to the complexity of the current cosmos?

63. How do the principles of holography and information theory inform our understanding of the universe as a vast interconnected network of information, where space, time, and matter are emergent phenomena?

64. Can we conceive of a multilevel model of reality that encompasses both the microcosmic realm of quantum mechanics and the macrocosmic

realm of relativity, bridging the gap between the quantum and classical worlds?

65. What role does the concept of emergence play in the evolution of complex systems, from the formation of galaxies to the emergence of life and consciousness, and how can we explain the phenomenon of novelty and creativity in the universe?

66. How do the principles of fractal geometry and self-similarity provide information about the structure and organization of the universe at all scales, from the subatomic to the cosmic?

67. Can we develop a coherent framework for understanding the relationship between mind and body that incorporates ideas from disciplines such as neuroscience, psychology, philosophy, and contemplative traditions?

68. How do the principles of symmetry and symmetry breaking in physics inform our understanding of the fundamental forces and particles of nature, and what role do symmetries play in the emergence of complexity and diversity in the universe?

69. Can we conceive of alternative modes of existence beyond the limitations of space-time, where concepts such as causality and locality may not apply, and if so, how might they manifest?

Quantum Metaphysical Investigations : Exploring the Depths of Reality and Consciousness
Marino Baca-Carmona

70. How do the principles of self-organization and criticality in complex systems illuminate the dynamics of consciousness and the emergence of coherent patterns of thought and perception within the human mind?

71. Can we develop a unified theory of consciousness that reconciles individuals' subjective experiences with objective measurements of brain activity, bridging the gap between first-person phenomenology and third-person neuroscience?

72. How do the principles of quantum field theory and string theory inform our understanding of the fundamental nature of reality, particularly in relation to the concept of a multiverse containing an infinite number of parallel universes?

73. Can we conceive of a meta-level of reality beyond the constraints of physical laws, where concepts like causality and locality may not apply, and if so, how might this meta-reality intersect or influence our own?

74. How do the principles of epigenetics and gene-environment interactions shape our understanding of the interaction between nature and nurture in the development of consciousness and identity?

75. What role does the concept of emergence play in the evolution of complex systems, from the formation of galaxies to the emergence of consciousness, and how can we reconcile the apparent teleology of emergent phenomena with the underlying principles of causality and determinism?

Quantum Metaphysical Investigations : <u>Exploring the Depths of Reality and Consciousness</u>
Marino Baca-Carmona

76. Can we develop a comprehensive theory of value that transcends subjective preferences and cultural biases, providing a universal framework for evaluating the value of actions, experiences, and entities in the universe?

77. How do the principles of computational complexity and algorithmic information theory shed light on the nature of consciousness as a computational process, and what implications do they have for our understanding of artificial intelligence and the possibility of conscious machines?

78. Can we conceive of alternative modes of perception and cognition that transcend the limitations of human senses and cultural conditioning, potentially leading to new ideas about the nature of reality and the fabric of the universe?

79. How do the principles of panpsychism and idealism inform our understanding of the relationship between mind and matter, and what implications do they have for our conceptions of consciousness, identity, and reality?

80. Can we develop a unified framework for understanding the origins and evolution of life, consciousness, and intelligence in the universe, incorporating ideas from fields such as astrobiology, complexity theory, and evolutionary psychology?

81. How do the principles of quantum cognition and Bayesian inference shed light on the nature of decision-making processes in the human

brain, and what implications do they have for our understanding of free will and determinism?

82. Can we develop a comprehensive theory of self that transcends the limitations of ego-centered identity and incorporates ideas from Eastern philosophies such as Advaita Vedanta and Buddhism?

83. How do the principles of nonlocality and entanglement in quantum mechanics challenge our notions of separation and individuality, and what implications do they have for our understanding of interconnectedness and unity in the universe?

84. Can we conceive of a metaethical framework that transcends cultural relativism and provides a universal basis for moral and ethical judgments, incorporating ideas from fields such as evolutionary psychology, game theory, and deontological ethics?

85. How do the principles of computational irreducibility and algorithmic randomness inform our limits on the limits of scientific prediction and the role of randomness and unpredictability in the universe?

86. Can we develop a coherent framework for understanding the relationship between consciousness and the physical brain that incorporates insights from fields such as quantum biology, neurophenomenology, and integrated information theory?

Quantum Metaphysical Investigations : Exploring the Depths of Reality and Consciousness
Marino Baca-Carmona

87. How do the principles of nonlinear dynamics and chaos theory inform our understanding of the emergence of complex patterns and structures in the universe, from the formation of galaxies to the evolution of consciousness?

88. Can we conceive of alternative modes of perception and cognition that transcend the limitations of human senses and cognitive biases, potentially leading to new insights into the nature of reality and the structure of the cosmos?

89. How do the principles of symmetry and symmetry breaking in physics inform our understanding of the origin and evolution of the fundamental forces and particles of nature, and what role do symmetries play in the emergence of complexity and diversity in physics? universe?

90. Can we develop a comprehensive theory of value that transcends subjective preferences and cultural prejudices, providing a universal basis for evaluating the value of actions, experiences, and entities in the cosmos?

91. How do the principles of quantum entanglement and nonlocality challenge our understanding of causality and the arrow of time, and what implications do they have for our conception of the past, present, and future?

92. Can we develop a unified framework for understanding the relationship between mind, matter, and consciousness that incorporates

Quantum Metaphysical Investigations : Exploring the Depths of Reality and Consciousness
Marino Baca-Carmona

ideas from both Eastern philosophies and modern scientific theories, such as panpsychism and quantum theories of mind?

93. How do the principles of quantum decoherence and wave function collapse inform our understanding of the transition from quantum uncertainty to classical certainty, and what role does measurement play in the manifestation of reality?

94. Can we conceive of a meta-level of reality beyond the limitations of space and time, where concepts such as identity and causality may not apply, and if so, how might this meta-reality interact or influence our own?

95. How do the principles of emergent phenomena and self-organization give rise to complex systems such as the human brain, societies, and ecosystems, and what insights can they provide about the nature of order and chaos in the universe?

96. Can we develop a comprehensive theory of aesthetics that transcends cultural relativism and provides a universal basis for appreciating beauty and artistic expression, incorporating ideas from fields such as psychology, neuroscience, and anthropology?

97. How do the principles of top-down causality and downward causation inform our understanding of the relationship between higher-level phenomena, such as consciousness and cognition, and their underlying physical substrates?

Quantum Metaphysical Investigations : Exploring the Depths of Reality and Consciousness

Marino Baca-Carmona

98. Can we conceive of alternative modes of existence beyond the limitations of space-time, where concepts such as causality and locality may not apply, and if so, how might they manifest and what implications do they have for our understanding of reality?

99. How do the principles of supersymmetry and string theory inform our understanding of the fundamental nature of reality, and what implications do they have for the existence of hidden dimensions and parallel universes?

100. Can we develop a coherent framework for understanding the role of intentionality and agency in the universe, incorporating ideas from fields such as quantum mechanics, philosophy of mind, and existential phenomenology?

101. How do the concept of cosmic inflation and the multiverse hypothesis influence our understanding of the nature of existence, particularly in relation to the emergence of universes with different physical laws and constants?

102. Can we conceive of a meta-level of reality beyond the limitations of causality and determinism, where concepts such as randomness and unpredictability govern the fabric of existence, and if so, how might this meta-reality intersect or influence ours?

103. How do the principles of hypercomputing and hypothetical models like Turing's oracle inform our understanding of the limits of

Quantum Metaphysical Investigations : Exploring the Depths of Reality and Consciousness
Marino Baca-Carmona

computability and the potential existence of entities capable of solving undecidable problems?

104. Can we develop a unified framework for understanding the relationship between consciousness and reality that incorporates ideas from quantum mechanics, information theory, and Eastern philosophies such as Advaita Vedanta and Taoism?

105. How do the concept of cosmic adjustment and the anthropic principle shape our understanding of the apparent adjustment of the fundamental constants and parameters of the universe for the emergence of life and consciousness?

106. Can we conceive of alternative modes of perception and cognition that transcend the limitations of human consciousness, potentially leading to new ideas about the nature of reality and the structure of existence?

107. How do the principles of nonduality and ego dissolution in mystical traditions such as Zen Buddhism and Advaita Vedanta inform our understanding of the nature of the self and reality?

108. Can we develop a comprehensive theory of meaning that transcends linguistic and cultural boundaries, providing a universal basis for understanding the nature of meaning and purpose in the universe?

109. How does the concept of retrocausality and time-symmetric interpretations of quantum mechanics challenge our understanding of

Quantum Metaphysical Investigations : Exploring the Depths of Reality and Consciousness
Marino Baca-Carmona

causality and the arrow of time, and what implications do they have for our conception of the past, present, and future?

110. Can we develop a coherent framework for understanding the relationship between mind and matter that incorporates ideas from fields such as panpsychism, quantum theories of mind, and philosophy of mind, providing a unified theory of consciousness and existence? ?

111. How do the principles of quantum gravity and the holographic principle challenge our understanding of the nature of space, time and gravity, and what implications do they have for our conception of reality as a holographic projection from a lower dimensional boundary?

112. Can we develop a unified framework for understanding the relationship between consciousness and the fundamental structure of the universe, incorporating ideas from fields such as quantum information theory, neurobiology, and cosmology?

113. How do the concept of eternal inflation and the landscape of string theory influence our understanding of the nature of existence, particularly in relation to the proliferation of universes with different properties and laws?

114. Can we conceive of a meta-level of reality beyond the limitations of causality and determinism, where concepts such as agency and intentionality govern the development of events, and if so, how might this meta-reality interact or influence ours?

Quantum Metaphysical Investigations : Exploring the Depths of Reality and Consciousness
Marino Baca-Carmona

115. How do the principles of quantum coherence and entanglement in biological systems inform our understanding of consciousness and cognition, and what implications do they have for our conception of mind-brain interactions and the nature of subjective experience?

116. Can we develop a comprehensive theory of cosmic consciousness that explains the interconnectedness of all beings and the universe as a single integrated entity, incorporating ideas from fields such as panpsychism, theories of the quantum mind, and transpersonal psychology?

117. How do the concept of cosmic inflation and the multiverse hypothesis influence our understanding of the origin and evolution of the universe, and what implications do they have for the existence of parallel realities and alternative histories?

118. Can we conceive of alternative modes of existence beyond the limitations of space-time, where concepts such as identity and causality may not apply, and if so, how might they manifest and what implications do they have for our understanding of reality?

119. How do principles of complexity and emergence in biological systems inform our understanding of the nature of life and consciousness, and what implications do they have for our conception of identity and purpose in the universe?

120. Can we develop a coherent framework for understanding the relationship between mind and matter that transcends dualistic notions of

Quantum Metaphysical Investigations : Exploring the Depths of Reality and Consciousness
Marino Baca-Carmona

consciousness and materialism, providing a unified theory of existence that encompasses both subjective experience and objective reality?

121. How does the concept of ontological pluralism and modal realism challenge our understanding of the nature of reality, particularly in relation to the existence of possible worlds and the multiverse hypothesis?

122. Can we conceive of a meta-level of reality beyond the limitations of logic and mathematics, where concepts such as paradox and contradiction are fundamental aspects of existence, and if so, how might this meta-reality manifest?

123. How do the principles of quantum field theory and the holographic principle inform our understanding of the nature of space, time and matter, and what implications do they have for our conception of reality as a holographic projection from a boundary lower dimension?

124. Can we develop a unified framework for understanding the relationship between consciousness and the fundamental structure of the universe, incorporating insights from fields such as quantum gravity, neurophenomenology, and transpersonal psychology?

125. How do the concept of cosmic fine-tuning and the anthropic principle shape our understanding of the apparent fine-tuning of the fundamental constants and parameters of the universe for the emergence of life and consciousness, and what implications do they have for our conception of purpose in the universe? cosmos?

Quantum Metaphysical Investigations : Exploring the Depths of Reality and Consciousness
Marino Baca-Carmona

126. Can we conceive of alternative modes of perception and cognition that transcend the limitations of human consciousness, potentially leading to new ideas about the nature of reality and the structure of existence beyond the comprehension of ordinary minds?

127. How do the principles of quantum information theory and the holographic principle inform our understanding of the nature of information and its role in shaping the fabric of reality, particularly in relation to the concept of the universe as a vast computer quantum?

128. Can we develop a comprehensive theory of cosmic consciousness that explains the interconnectedness of all beings and the universe as a single integrated entity, incorporating ideas from fields such as panpsychism, transpersonal psychology, and mysticism?

129. How do the concept of eternal inflation and the landscape of string theory influence our understanding of the nature of existence, particularly in relation to the proliferation of universes with different properties and laws, and what implications do they have for our conception of existence? reality as a multiverse?

130. Can we conceive of a meta-level of reality beyond the constraints of space-time and causality, where concepts such as existence and non-existence are intertwined in a higher-dimensional tapestry of being, and if so, how might this meta-reality challenge our understanding of the nature of existence itself?

Quantum Metaphysical Investigations : Exploring the Depths of Reality and Consciousness
Marino Baca-Carmona

131. How does the concept of quantum entanglement and non-locality challenge our understanding of the nature of reality, particularly in relation to the interconnectedness of all phenomena and the potential existence of hidden variables governing quantum systems?

132. Can we conceive of a meta-level of reality beyond the limitations of language and communication, where concepts such as meaning and interpretation are transcended by a higher-dimensional mode of understanding, and if so, how might this meta- reality shape our perception of existence?

133. How do principles of hypercomputability and hypothetical models such as Turing's oracle inform our understanding of the limits of computation and the potential existence of entities capable of solving undecided problems beyond the capabilities of classical computers?

134. Can we develop a unified framework for understanding the relationship between consciousness and the underlying fabric of reality, incorporating insights from fields such as quantum gravity, integrated information theory, and transpersonal psychology?

135. How do the concept of cosmic inflation and the multiverse hypothesis influence our understanding of the nature of existence, particularly in relation to the emergence of universes with different physical laws and constants, and what implications do they have for our conception of reality? like a vast and evolving cosmic landscape?

Quantum Metaphysical Investigations : <u>Exploring the Depths of Reality and Consciousness</u>
Marino Baca-Carmona

136. Can we conceive of alternative modes of perception and cognition that transcend the limitations of human consciousness, potentially leading to new ideas about the nature of reality and the structure of existence inaccessible to ordinary minds?

137. How do the principles of quantum cosmology and the wave function of the universe inform our understanding of the origins and evolution of the cosmos, particularly in relation to the concept of a quantum state of the entire universe that encompasses all possible histories and outcomes ?

138. Can we develop a comprehensive theory of cosmic consciousness that takes into account the interconnectedness of all beings and the universe as a single unified entity, incorporating ideas from fields such as panpsychism, theories of the quantum mind, and Eastern mysticism?

139. How do the concept of eternal recurrence and the cyclical nature of time influence our understanding of the nature of existence, particularly in relation to the repetition of events and patterns throughout cosmic history, and what implications do they have for our conception of free will and destiny?

140. Can we conceive of a meta-level of reality beyond the limitations of causality and determinism, where concepts such as agency and intentionality govern the development of events in a timeless and higher-dimensional continuum of existence, and if it is So how might this meta-reality shape our understanding of the fundamental nature of reality?

Quantum Metaphysical Investigations : Exploring the Depths of Reality and Consciousness
Marino Baca-Carmona

141. How does the concept of quantum superposition and wave function collapse challenge our understanding of the nature of reality, particularly in relation to the role of observation and measurement in shaping the outcomes of quantum phenomena?

142. Can we conceive of a meta-level of reality beyond the constraints of causality and temporality, where concepts such as causality and time are transcended by a higher-dimensional mode of existence, and if so, how could this meta-level -does reality influence our perception of the universe?

143. How do the principles of quantum gravity and the holographic principle inform our understanding of the nature of space, time and information, particularly in relation to the holographic encoding of reality in a lower dimensional boundary?

144. Can we develop a unified theory of consciousness that incorporates ideas from both Eastern contemplative traditions and modern neuroscience, providing a comprehensive understanding of the nature of consciousness and subjective experience?

145. How does the concept of cosmic symmetry breaking and phase transitions influence our understanding of the origin and evolution of the universe, particularly in relation to the emergence of structure and complexity from an initial state of symmetry?

146. Can we conceive of alternative modes of perception and cognition that transcend the limitations of human consciousness, potentially

Quantum Metaphysical Investigations : <u>Exploring the Depths of Reality and Consciousness</u>
Marino Baca-Carmona

leading to new ideas about the nature of reality and the interconnectedness of all things?

147. How do the principles of eternal inflation and the landscape of string theory inform our understanding of the vastness and diversity of the cosmos, particularly in relation to the existence of multitudes of parallel universes with different properties and laws?

148. Can we develop a coherent framework for understanding the relationship between mind and matter that transcends dualistic notions of consciousness and materialism, providing a unified theory of existence that encompasses both subjective experience and objective reality?

149. How do the concept of quantum decoherence and the emergence of classical behavior of quantum systems influence our understanding of the transition from the quantum realm to the classical world, particularly in relation to the problem of measurement and the role of observation?

150. Can we conceive of a meta-level of reality beyond the limitations of language and representation, where concepts such as meaning and representation are transcended by a higher-dimensional mode of understanding, and if so, how might this meta- reality shape our perception of existence?

151. How do the concept of quantum information encoding and the black hole information paradox challenge our understanding of the conservation of information and the nature of spacetime in the presence of gravitational singularities?

Quantum Metaphysical Investigations : Exploring the Depths of Reality and Consciousness
Marino Baca-Carmona

152. Can we conceive of a meta-level of reality beyond the limitations of causality and logic, where concepts such as causality and logic are transcended by a higher dimensional mode of existence, and if so, how could this meta-reality shape our understanding of the universe?

153. How do the principles of quantum entanglement and the holographic principle inform our understanding of the nature of entanglement entropy and the holographic encoding of information at the boundary of a black hole event horizon?

154. Can we develop a unified theory of consciousness that integrates insights from both Western neuroscience and Eastern meditative practices, providing a comprehensive understanding of the nature of consciousness and its relationship to the fabric of reality?

155. How do the concept of cosmic inflation and the multiverse hypothesis influence our understanding of the fundamental constants and parameters of the universe, particularly in relation to the existence of a vast set of parallel universes with different properties and laws?

156. Can we conceive of alternative modes of perception and cognition that transcend the limitations of human consciousness, potentially leading to new insights into the nature of reality and the underlying fabric of existence?

157. How do the principles of eternal inflation and the landscape of string theory inform our understanding of the vastness and complexity of

Quantum Metaphysical Investigations : Exploring the Depths of Reality and Consciousness
Marino Baca-Carmona

the multiverse, particularly in relation to the existence of universes with different dimensionality and space-time geometries?

158. Can we develop a coherent framework for understanding the relationship between mind and matter that transcends dualistic conceptions of consciousness and materialism, providing a unified theory of existence that encompasses both subjective experience and objective reality?

159. How does the concept of quantum teleportation and non-local communication challenge our understanding of the limits of locality and the nature of space-time as a causal framework for the propagation of information and interactions?

160. Can we conceive of a meta-level of reality beyond the limitations of language and representation, where concepts such as meaning and representation are transcended by a higher-dimensional mode of understanding, and if so, how might this meta- reality shape our perception of existence?

161. How do the concept of quantum consciousness and Orch-OR theory challenge our understanding of the relationship between quantum mechanics and the emergence of subjective experience, particularly in relation to the role of microtubules in neural processing?

162. Can we conceive of a meta-level of reality beyond the constraints of causality and locality, where concepts such as causality and space-time are transcended by a higher-dimensional mode of existence, and if

Quantum Metaphysical Investigations : <u>Exploring the Depths of Reality and Consciousness</u>
Marino Baca-Carmona

so, how might Will this meta-reality shape our perception of the cosmos?

163. How do the principles of cosmic inflation and brane cosmology inform our understanding of the landscape of the multiverse and the existence of parallel universes with different laws and physical constants?

164. Can we develop a unified theory of consciousness that incorporates ideas from both modern neuroscience and ancient contemplative practices, providing a comprehensive understanding of the nature of consciousness and its relationship to the underlying fabric of reality?

165. How do the concept of quantum gravity and the holographic principle challenge our understanding of space, time and gravity at the Planck scale, particularly in relation to the encoding of information at the boundary of a black hole event horizon?

166. Can we conceive of alternative modes of perception and cognition that transcend the limitations of human consciousness, potentially revealing new dimensions of reality and the interconnectedness of all things?

167. How do the principles of quantum entanglement and nonlocality inform our understanding of the nature of reality and the possibility of instantaneous communication across great distances?

Quantum Metaphysical Investigations : <u>Exploring the Depths of Reality and Consciousness</u>

Marino Baca-Carmona

168. Can we develop a coherent framework for understanding the relationship between mind and matter that transcends traditional dualistic conceptions, providing a unified theory of existence that encompasses both subjective experience and objective reality?

169. How does the concept of cosmic fine-tuning and the anthropic principle shape our understanding of the remarkable suitability of the universe for the emergence of life and consciousness?

170. Can we conceive of a meta-level of reality beyond the limitations of language and representation, where concepts such as meaning and representation are transcended by a higher-dimensional mode of understanding, and if so, how might this meta-level reality shape our perception of existence?

171. How do the concept of quantum superposition and the many-worlds interpretation challenge our understanding of the nature of reality, particularly in relation to the coexistence of multiple branching universes with different outcomes?

172. Can we conceive of a meta-level of reality beyond the limitations of causality and temporality, where concepts such as causality and time are transcended by a higher-dimensional mode of existence, and if so, how could this meta-level -reality influence our perception of the existence and development of events?

173. How do the principles of quantum entanglement and quantum teleportation inform our understanding of the nature of information and

communication, particularly in relation to the possibility of instantaneous transfer of quantum states over arbitrary distances?

174. Can we develop a unified theory of consciousness that incorporates ideas from both Eastern contemplative traditions and Western cognitive neuroscience, providing a comprehensive understanding of the nature of consciousness and its relationship to the underlying fabric of reality?

175. How do the concept of cosmic inflation and the multiverse hypothesis influence our understanding of the fundamental constants and parameters of the universe, particularly in relation to the existence of a set of parallel universes with different properties and laws?

176. Can we conceive of alternative modes of perception and cognition that transcend the limitations of human consciousness, potentially revealing new dimensions of reality and the interconnectedness of all things beyond our current understanding?

177. How do the principles of eternal inflation and the landscape of string theory inform our understanding of the vastness and complexity of the multiverse, particularly in relation to the existence of universes with different dimensionality and space-time geometries?

178. Can we develop a coherent framework for understanding the relationship between mind and matter that transcends dualistic conceptions, providing a unified theory of existence that encompasses both subjective experience and objective reality?

Quantum Metaphysical Investigations : Exploring the Depths of Reality and Consciousness

Marino Baca-Carmona

179. How does the concept of quantum computing and quantum algorithms challenge our understanding of computing and problem solving, particularly in relation to the potential for exponential acceleration and the solution of currently intractable problems?

180. Can we conceive of a meta-level of reality beyond the limitations of language and representation, where concepts such as meaning and representation are transcended by a higher-dimensional mode of understanding, and if so, how might this meta- reality shape our perception of the existence and nature of truth?

181. How do the concept of cosmic censorship and the firewall paradox challenge our understanding of the nature of black holes and the information paradox, particularly in relation to the preservation of information within the event horizon?

182. Can we conceive of a meta-level of reality beyond the limitations of dimensionality and geometry, where concepts such as spatial dimensions and geometric shapes are transcended by a higher dimensional mode of existence, and if so, how Could this meta-reality influence our perception of the cosmos?

183. How do the principles of quantum entanglement and quantum teleportation inform our understanding of the potential of quantum communication and quantum networks, particularly in relation to the secure transmission of information over quantum channels?

Quantum Metaphysical Investigations : Exploring the Depths of Reality and Consciousness

Marino Baca-Carmona

184. Can we develop a unified theory of consciousness that incorporates insights from both contemporary neuroscience and ancient wisdom traditions, providing a comprehensive understanding of the nature of consciousness and its role in shaping reality?

185. How does the concept of cosmic inflation and the eternal inflationary landscape inform our understanding of the existence of bubble universes and the possibility of regions of space-time with different physical laws and constants?

186. Can we conceive of alternative modes of perception and cognition that transcend the limitations of human consciousness, potentially revealing new realms of existence and the interconnectedness of all things beyond our current cognitive abilities?

187. How do the principles of emergent gravity and quantum information theory inform our understanding of the emergence of spacetime and gravity from the underlying quantum degrees of freedom, particularly in relation to the holographic nature of gravity?

188. Can we develop a coherent framework for understanding the relationship between mind and matter that transcends dualistic conceptions, providing a unified theory of existence that encompasses both subjective experience and objective reality in a quantum framework?

189. How does the concept of quantum error correction and fault-tolerant quantum computing challenge our understanding of the limits of

computation and the potential of error-free quantum computing in the presence of noise and decoherence?

190. Can we conceive of a meta-level of reality beyond the limitations of language and representation, where concepts such as meaning and representation are transcended by a higher-dimensional mode of understanding, and if so, how might this meta- actually shape our perception of the existence and nature of knowledge?

191. How does the concept of cosmic inflation and the eternal inflationary landscape inform our understanding of the origin and evolution of the universe, particularly in relation to the existence of bubble universes and the potential for a cyclical cosmological model?

192. Can we conceive of a meta-level of reality beyond the limitations of causality and determinism, where concepts such as causality and determinism are transcended by a higher dimensional mode of existence, and if so, how could this meta-reality shape our perception of the cosmos?

193. How do the principles of quantum gravity and loop quantum cosmology inform our understanding of the quantum nature of the early universe and the possibility of a quantum origin of space and time?

194. Can we develop a unified theory of consciousness that incorporates ideas from both Western neuroscience and Eastern contemplative traditions, providing a comprehensive understanding of the nature of consciousness and its relationship to the fabric of reality?

Quantum Metaphysical Investigations : Exploring the Depths of Reality and Consciousness
Marino Baca-Carmona

195. How do the concept of cosmic fine-tuning and the anthropic principle shape our understanding of the fundamental constants and parameters of the universe, particularly in relation to the existence of a vast multiverse with different properties and laws?

196. Can we conceive of alternative modes of perception and cognition that transcend the limitations of human consciousness, potentially revealing new dimensions of reality and the interconnectedness of all things beyond our current cognitive abilities?

197. How do the principles of eternal inflation and the landscape of string theory inform our understanding of the vastness and complexity of the multiverse, particularly in relation to the existence of universes with different dimensionality and space-time geometries?

198. Can we develop a coherent framework for understanding the relationship between mind and matter that transcends traditional dualistic conceptions, providing a unified theory of existence that encompasses both subjective experience and objective reality?

199. How does the concept of quantum computing and quantum algorithms challenge our understanding of computing and problem solving, particularly in relation to the potential for exponential acceleration and the solution of currently intractable problems?

200. Can we conceive of a meta-level of reality beyond the limitations of language and representation, where concepts such as meaning and

Quantum Metaphysical Investigations : Exploring the Depths of Reality and Consciousness
Marino Baca-Carmona

representation are transcended by a higher-dimensional mode of understanding, and if so, how might this meta-level reality shape our perception of the existence and nature of truth?

201. How do the concept of quantum entanglement and its potential for nonlocal correlations challenge our understanding of locality and the nature of reality, particularly in relation to the EPR paradox and Bell's theorem?

202. Can we conceive of a meta-level of reality beyond the constraints of space-time and causality, where concepts such as causality and temporal order are transcended by a higher-dimensional mode of existence, and if so, how might Does this meta-reality influence our perception of the universe?

203. How do the principles of quantum gravity and holographic duality inform our understanding of the fundamental structure of space-time and the encoding of information at its boundary, particularly in relation to the holographic nature of black holes and the universe?

204. Can we develop a unified theory of consciousness that integrates ideas from both Eastern contemplative traditions and Western neuroscience, providing a comprehensive understanding of the nature of subjective experience and its relationship to the underlying fabric of reality?

205. How do the concept of cosmic inflation and the multiverse hypothesis influence our understanding of the origins and evolution of

Quantum Metaphysical Investigations : Exploring the Depths of Reality and Consciousness
Marino Baca-Carmona

the universe, particularly in relation to the possibility of eternal inflation and the existence of a landscape of vacuum states?

206. Can we conceive of alternative modes of perception and cognition that transcend the limitations of human consciousness, potentially revealing new dimensions of reality and the interconnectedness of all things beyond our current cognitive abilities?

207. How do the principles of quantum information theory and quantum computing inform our understanding of the computational nature of reality and the potential of quantum algorithms to solve classically intractable problems?

208. Can we develop a coherent framework for understanding the relationship between mind and matter that transcends traditional dualistic conceptions, providing a unified theory of existence that encompasses both subjective experience and objective reality?

209. How do the concept of quantum entanglement and its role in quantum communication challenge our understanding of information transmission and the nature of correlations between distant particles, particularly in relation to the implications for secure quantum communication protocols?

210. Can we conceive of a meta-level of reality beyond the limitations of language and representation, where concepts such as meaning and representation are transcended by a higher-dimensional mode of

Quantum Metaphysical Investigations : <u>Exploring the Depths of Reality and Consciousness</u>
Marino Baca-Carmona

understanding, and if so, how might this meta- actually shape our perception of the existence and nature of knowledge?

211. How do the concept of cosmic holography and the holographic principle challenge our understanding of the nature of reality, particularly in relation to the encoding of information on lower dimensional surfaces and its projection into three-dimensional space?

212. Can we conceive of a meta-level of reality beyond the limitations of physicality and form, where concepts such as substance and structure are transcended by a higher-dimensional mode of existence, and if so, how could this meta-level -Does reality influence our perception of existence and the fabric of the cosmos?

213. How do the principles of quantum cosmology and the wave function of the universe inform our understanding of the quantum origins of the cosmos and the potential for an autonomous quantum reality without the need for external causality?

214. Can we develop a unified theory of consciousness that integrates insights from both objective neuroscience and subjective phenomenology, providing a comprehensive understanding of the nature of subjective experience and its relationship to the underlying quantum fabric of reality?

215. How do the concept of cosmic fine-tuning and the multiverse landscape challenge our understanding of the fundamental parameters of

the universe, particularly in relation to the possibility of a finely tuned set of universes with different physical laws and constants?

216. Can we conceive of alternative modes of perception and cognition that transcend the limitations of human consciousness, potentially revealing new dimensions of reality and the interconnectedness of all things beyond our current cognitive abilities?

217. How do the principles of emergent gravity and quantum entanglement inform our understanding of the emergence of spacetime and gravity from the underlying quantum degrees of freedom, particularly in relation to the holographic nature of gravity?

218. Can we develop a coherent framework for understanding the relationship between mind and matter that transcends traditional dualistic conceptions, providing a unified theory of existence that encompasses both subjective experience and objective reality?

219. How does the concept of quantum computing and quantum complexity challenge our understanding of computing and problem solving, particularly in relation to the potential of quantum systems to process information in ways that classical computers cannot?

220. Can we conceive of a meta-level of reality beyond the limitations of language and representation, where concepts such as meaning and representation are transcended by a higher-dimensional mode of understanding, and if so, how might this meta- actually shape our perception of the existence and nature of knowledge?

Quantum Metaphysical Investigations : Exploring the Depths of Reality and Consciousness
Marino Baca-Carmona

221. How do the concept of quantum immortality and the many-worlds interpretation challenge our understanding of mortality and the nature of consciousness, particularly in relation to the subjective experience of eternal survival across branching timelines?

222. Can we conceive of a meta-level of reality beyond the limitations of dimensionality and topology, where concepts such as spatial dimensions and topological properties are transcended by a higher-dimensional mode of existence, and if so, how might Will this meta-reality shape our perception of the universe?

223. How do the principles of quantum cosmology and the multiverse landscape inform our understanding of the origin and evolution of the cosmos, particularly in relation to the existence of a set of universes with different initial conditions and evolutionary trajectories?

224. Can we develop a unified theory of consciousness that incorporates ideas from both Western neurobiology and Eastern spiritual traditions, providing a comprehensive understanding of the nature of consciousness and its relationship to the underlying fabric of reality?

225. How does the concept of cosmic tuning and the anthropic principle shape our understanding of the fundamental parameters of the universe, particularly in relation to the existence of a finely tuned multiverse leading to the emergence of life and consciousness?

Quantum Metaphysical Investigations : Exploring the Depths of Reality and Consciousness

Marino Baca-Carmona

226. Can we conceive of alternative modes of perception and cognition that transcend the limitations of human consciousness, potentially revealing new dimensions of reality and the interconnectedness of all things beyond our current cognitive abilities?

227. How do the principles of emergent spacetime and holographic duality inform our understanding of the emergence of spacetime and gravity from the underlying quantum degrees of freedom, particularly in relation to the holographic encoding of information about the limit of black holes?

228. Can we develop a coherent framework for understanding the relationship between mind and matter that transcends traditional dualistic conceptions, providing a unified theory of existence that encompasses both subjective experience and objective reality?

229. How does the concept of quantum computing and quantum algorithms challenge our understanding of computing and problem solving, particularly in relation to the potential of quantum systems to solve classically intractable problems and simulate complex physical systems?

230. Can we conceive of a meta-level of reality beyond the limitations of language and representation, where concepts such as meaning and representation are transcended by a higher-dimensional mode of understanding, and if so, how might this meta-level actually shape our perception of the existence and nature of knowledge?

Quantum Metaphysical Investigations : Exploring the Depths of Reality and Consciousness
Marino Baca-Carmona

231. In a hypothetical scenario in which the observer effect extends beyond the quantum realm into macroscopic phenomena, how would the act of observation itself alter the fabric of reality, and what implications would this have for our understanding of causality and determinism?

232. Can we imagine a meta-level of reality in which consciousness itself is the fundamental building block, giving rise to both the observer and the observed, and if so, how might this consciousness-centered perspective reshape our understanding of reality? the nature of existence?

233. How do the principles of quantum entanglement and non-locality intersect with the notion of cosmic interconnectedness, and could this interconnectedness extend beyond spatial and temporal boundaries, uniting all entities into a vast cosmic web of entangled consciousness?

234. Can we conceive of a reality in which time is not simply a linear progression, but rather a multidimensional construct, where the past, present and future coexist simultaneously, and if so, how might this timeless reality redefine our Understanding causality and free will?

235. How could the principles of superposition and quantum decoherence be applied not only to particles, but also to entire universes, leading to the coexistence of multiple parallel realities with different histories and outcomes, and what role would conscious observation play in the collapse of these quantum possibilities?

236. Can we develop a comprehensive theory of consciousness that transcends reductionist approaches and integrates ideas from diverse disciplines such as quantum physics, neuroscience, philosophy, and mysticism, providing a unified framework for understanding the nature of subjective experience?

237. In a universe governed by the principles of quantum cosmology, where the wave function of the universe encompasses all possible states, how would the concept of cosmic self-awareness manifest and what implications would this have for our perception of the cosmos as a conscious entity? ?

238. How might the concept of cosmic intelligence, as proposed in theories such as panpsychism and cosmopsychism, influence our understanding of the purpose and meaning of existence, and what role would sentient beings play in the cosmos' evolutionary journey toward self-awareness?

239. Can we conceive of a reality in which the boundaries between the physical world and the realm of consciousness become blurred, leading to the emergence of phenomena such as psychokinesis and mind-matter interaction, and how this paradigm shift would challenge our conventional scientific worldview?

240. How might the principles of quantum complexity and computational irreducibility shape the dynamics of cosmic evolution, leading to the emergence of self-organizing structures and the

Quantum Metaphysical Investigations : <u>Exploring the Depths of Reality and Consciousness</u>
Marino Baca-Carmona

spontaneous emergence of consciousness at various levels of cosmic complexity?

241. In a universe where fundamental constants vary across different regions of space-time, how would the existence of "pocket universes" with radically different physical laws challenge our understanding of cosmology and the nature of reality?

242. Can we conceive of a reality in which consciousness itself is the fabric of existence, with the entire cosmos arising from one primordial consciousness, and if so, how might this non-dualistic perspective reshape our understanding of ontology and nature? of being?

243. How do the principles of quantum cognition and Bayesian brain theory intersect with the concept of the "quantum mind," and what implications does this have for our understanding of perception, decision-making, and the nature of reality?

244. Can we develop a unified theory of consciousness that incorporates ideas from both Western empirical science and Eastern contemplative traditions, providing a comprehensive framework for understanding the nature of subjective experience and its relationship to the cosmos?

245. In a multiverse scenario where each universe is governed by different laws of physics, how would the existence of the "parent" and "child" universes influence the concept of causality, and could this lead to a perpetual cycle? of creation and destruction?

Quantum Metaphysical Investigations : Exploring the Depths of Reality and Consciousness
Marino Baca-Carmona

246. How might the concept of cosmic entanglement extend beyond the quantum realm, linking disparate regions of the universe into a web of interconnection, and what role would conscious observers play in the collapse of these cosmic entanglements?

247. Can we conceive of a reality in which time is not a linear progression, but rather a complex network of causal loops and temporal paradoxes, and how would this model of "non-linear time" be reconciled with our everyday experience of time?

248. How do the principles of quantum gravity and loop quantum cosmology inform our understanding of the origins of the universe and the possibility of a pre-Big Bang epoch where the universe undergoes cycles of expansion and contraction?

249. Can we develop a mathematical framework that encompasses both quantum mechanics and general relativity, providing a unified theory of quantum gravity that resolves the singularities of black holes and the beginning of the universe?

250. In a universe where consciousness is not limited to biological entities, but permeates all levels of reality, how would this "panpsychic cosmos" manifest and what implications would it have for our understanding of the nature of reality and the role of conscious observers in its configuration?

251. Can we imagine a reality in which the concept of self is an illusion, and individual consciousnesses are simply transient manifestations of a

Quantum Metaphysical Investigations : <u>Exploring the Depths of Reality and Consciousness</u>
Marino Baca-Carmona

larger cosmic consciousness, and if so, how would this paradigm shift redefine our understanding of identity? and existence?

252. How could the concept of quantum entanglement be extended beyond particles to encompass entire systems and even the universe as a whole, leading to the emergence of a "cosmic entanglement" that transcends spatial and temporal boundaries?

253. Can we conceive of a reality in which the laws of physics themselves are emergent properties arising from a deeper underlying structure, and if so, what implications would this have for our understanding of causality and the nature of reality?

254. In a universe where time is not a linear progression, but rather a multidimensional construct, how would the concept of "time loops" and causal loops shape the dynamics of cosmic evolution, and could this lead to a self-referential universe that is perpetually recreated?

255. How do the principles of quantum information theory and the holographic principle inform our understanding of the universe as a vast information processing system, and what implications does this have for our perception of reality as a holographic projection?

256. Can we develop a comprehensive theory of consciousness that incorporates ideas from diverse disciplines such as physics, biology, psychology, and philosophy, providing a unified framework for understanding the nature of subjective experience and its relationship to the cosmos?

Quantum Metaphysical Investigations : Exploring the Depths of Reality and Consciousness

Marino Baca-Carmona

257. In a reality where the boundaries between the physical and the mental are blurred, how would the concept of "mind over matter" manifest, and could conscious intentionality influence the fabric of reality at the quantum level?

258. How might the existence of higher-dimensional branes and compacted dimensions shape the topology of the universe, leading to the emergence of exotic phenomena such as wormholes, extradimensional beings, and parallel realities?

259. Can we conceive of a universe where the same laws of physics are subject to evolution and adaptation, leading to the emergence of new physical constants, particles and forces over cosmic time, and what role would conscious observers play in this process? evolutionary?

260. In a reality where each observation collapses the wave function into a specific outcome, how would the concept of "observer-created reality" manifest and what implications would this have for our understanding of free will, determinism and the nature of existence itself?

261. Can we conceive of a reality in which the very notion of existence is probabilistic, with entities fluctuating in and out of existence based on quantum probabilities, and if so, how would this probabilistic ontology redefine our understanding of reality?

262. How might the concept of cosmic consciousness manifest in a universe where every sentient being is a node in a vast interconnected

network of consciousness, and what implications would this have for our understanding of individuality and identity?

263. Can we develop a mathematical framework that encompasses both quantum mechanics and general relativity, while incorporating higher-dimensional structures, providing a unified theory of quantum gravity and brane cosmology?

264. In a universe where time is not a fundamental aspect of reality, but rather an emergent property, how would the concept of "timeless consciousness" manifest and what implications would this have for our understanding of the nature of subjective experience? ?

265. How might the principles of quantum cognition and Bayesian brain theory inform our understanding of consciousness as a probabilistic process, with perceptions and decisions arising from a complex interplay of quantum uncertainties?

266. Can we conceive of a reality in which consciousness itself is the fundamental substratum of existence, with matter and energy arising as secondary manifestations of conscious consciousness, and if so, how does this "consciousness-centered" ontology Would it reshape our understanding of the cosmos?

267. How do the principles of cosmic inflation and eternal inflation inform our understanding of the multiverse as a vast set of bubble universes, each with its own set of physical laws and constants, and what implications does this have for the concept of adjustment cosmic fine?

Quantum Metaphysical Investigations : Exploring the Depths of Reality and Consciousness
Marino Baca-Carmona

268. In a universe governed by the principles of quantum cosmology, where the wave function of the universe encompasses all possible states, how would the concept of "cosmic self-awareness" manifest and what role would sentient beings play in this self-referential cosmic consciousness? ?

269. Can we develop a unified theory of consciousness that integrates knowledge from diverse fields such as physics, biology, psychology and philosophy, providing a comprehensive understanding of the nature of subjective experience and its relationship to the underlying fabric of consciousness? reality?

270. In a reality where each observation collapses the wave function into a specific outcome, how would the concept of "observer-induced reality" manifest, and what implications would this have for our understanding of the nature of perception? , reality and the observed relationship between the observer?

271. Can we conceive of a reality in which the laws of physics themselves are emergent phenomena, arising from a deeper underlying structure or substrate, and if so, how might this "physics from the ground up" perspective reshape our understanding of the nature of reality?

272. How might the principles of quantum cognition and the Bayesian brain hypothesis intersect with the concept of a "quantum mind," where consciousness arises from quantum interactions within the brain, and

Quantum Metaphysical Investigations : <u>Exploring the Depths of Reality and Consciousness</u>
Marino Baca-Carmona

what implications does this have for our understanding of consciousness? perception and decision making?

273. In a universe where every conceivable outcome exists simultaneously within the quantum wave function, how would the concept of "quantum branching" manifest and what role would conscious observation play in selecting and navigating between these branching timelines?

274. Can we develop a comprehensive theory of consciousness that incorporates ideas from both empirical neuroscience and subjective phenomenology, providing a unified framework for understanding the nature of subjective experience and its relationship to the underlying fabric of reality?

275. How might the concept of "ontological emergence" shape our understanding of the emergence of complex phenomena such as consciousness, where novel properties and behaviors emerge from the interactions of simpler elements, and what role would quantum phenomena play in this emerging process?

276. In a universe where the concept of locality breaks down to the quantum level, how would the existence of "quantum nonlocality" manifest on cosmic scales, and what implications would this have for our understanding of causality and the nature of reality? ?

277. Can we conceive of a reality in which consciousness itself is the fundamental building block of the universe, giving rise to both the

Quantum Metaphysical Investigations : Exploring the Depths of Reality and Consciousness
Marino Baca-Carmona

observer and the observed, and if so, how might this "consciousness-centered" ontology reshape our understanding of the universe? cosmos?

278. How do the principles of cosmic inflation and brane cosmology inform our understanding of the multiverse as a vast set of bubble universes, each with their own unique properties and evolutionary trajectories, and what implications does this have for our concept of cosmic identity? ?

279. In a universe where the concept of time is fluid and malleable, how would the existence of "time loops" and temporal paradoxes shape the dynamics of cosmic evolution, and could this lead to a cyclical model of the cosmos where the Does history repeat itself indefinitely?

280. Can we develop a mathematical framework that encompasses both quantum mechanics and general relativity, while incorporating the concept of consciousness as a fundamental aspect of reality, providing a unified theory of quantum gravity and conscious cosmology?

281. In a universe where the fabric of spacetime itself is subject to quantum fluctuations, how could these fluctuations give rise to emergent phenomena such as gravitational waves, quantum foam, and spacetime singularities, and what implications would they have? this for our understanding of the nature of reality?

282. Can we conceive of a reality in which the laws of physics are not fixed, but rather evolve over cosmic time, leading to the emergence of

Quantum Metaphysical Investigations : <u>Exploring the Depths of Reality and Consciousness</u>
Marino Baca-Carmona

new physical phenomena and structures, and what role would conscious observers play in this continuous process of cosmic evolution?

283. How could the principles of quantum entanglement and nonlocality extend beyond the realm of particles to encompass entire systems and even the universe as a whole, leading to the emergence of a "cosmic entanglement" that transcends spatial and temporal limits?

284. In a universe where all possible outcomes of a quantum event are realized simultaneously within the quantum wave function, how would the concept of "quantum superposition" manifest on cosmic scales, and what implications would this have for our understanding of cosmic structure and dynamics?

285. Can we develop a unified theory of consciousness that incorporates ideas from both empirical neuroscience and transpersonal psychology, providing a comprehensive framework for understanding the nature of subjective experience and its relationship to the cosmos as a whole?

286. How might the concept of cosmic consciousness manifest in a universe where every sentient being is a node in a vast interconnected network of consciousness, and what implications would this have for our understanding of identity, agency, and the nature of existence?

287. In a reality where the boundaries between the physical and the mental are blurred, how would the concept of "mind over matter" manifest and what implications would this have for our understanding of

Quantum Metaphysical Investigations : <u>Exploring the Depths of Reality and Consciousness</u>
Marino Baca-Carmona

the nature of reality and the role of conscious intentionality in its configuration?

288. Can we conceive of a universe where the concept of causality is not absolute, but is subject to revision and reevaluation, leading to the emergence of phenomena such as retrocausality and causal loops, and what implications would this have for our concept of causality? time and causality?

289. How do the principles of quantum cognition and the Bayesian brain hypothesis inform our understanding of decision making, perception, and the nature of consciousness, particularly in relation to the probabilistic nature of quantum phenomena?

290. In a reality where the boundaries between individual minds are porous and interconnected, how might the concept of "collective consciousness" manifest and what role would conscious intentionality play in shaping the dynamics of this collective consciousness?

291. Can we imagine a reality in which the laws of physics themselves evolve over cosmic time, leading to the emergence of new physical phenomena and structures, and what role would conscious observers play in this continuous process of evolution? cosmic?

292. How might the principles of quantum entanglement and nonlocality extend beyond the realm of particles to encompass entire systems and even the fabric of spacetime itself, leading to the emergence of

Quantum Metaphysical Investigations : Exploring the Depths of Reality and Consciousness
Marino Baca-Carmona

"quantum interconnectedness"? that transcends conventional notions of locality and separation?

293. In a universe where the concept of time is not a linear progression, but rather a multidimensional construct, how would the concept of "temporal entanglement" manifest and what implications would this have for our understanding of causality, free will and the nature of existence?

294. Can we develop a unified theory of consciousness that incorporates insights from both empirical neuroscience and Eastern contemplative traditions, providing a comprehensive framework for understanding the nature of subjective experience and its relationship to the fabric of reality?

295. How might the concept of cosmic consciousness manifest in a universe where each sentient being is a node in a vast interconnected network of consciousness, and what implications would this have for our understanding of identity, agency, and the nature of existence?

296. In a reality where the boundaries between the physical and the mental are blurred, how would the concept of "conscious matter" manifest and what role would conscious intentionality play in shaping the dynamics of this conscious material world?

297. Can we conceive of a universe in which the very fabric of space-time is imbued with consciousness, giving rise to a "cosmic mind" that

Quantum Metaphysical Investigations : Exploring the Depths of Reality and Consciousness
Marino Baca-Carmona

permeates all levels of reality, and if so, how could this cosmic consciousness influence the development of cosmic events?

298. How do the principles of quantum cognition and the Bayesian brain hypothesis inform our understanding of decision making, perception, and the nature of consciousness, particularly in relation to the probabilistic nature of quantum phenomena?

299. In a reality where each observation collapses the wave function into a specific outcome, how would the concept of "observer-induced reality" manifest, and what implications would this have for our understanding of the nature of perception? , reality and the observed relationship between the observer?

300. Can we conceive of a universe where the very concept of existence itself is fluid and mutable, with entities fluctuating in and out of existence based on the probabilistic nature of quantum reality, and what implications would this have for our understanding of existence? ontology and the nature of being?

Answers

1. The evolving laws of physics within a cosmic context point to the dynamic nature of the universe and its continuous search for balance amidst an ever-changing landscape. This evolution is not arbitrary, but is guided by fundamental principles, such as symmetry, conservation, and emergent complexity. Through cosmic epochs, the universe undergoes phase transitions, symmetry breaks and self-organizing processes,

Quantum Metaphysical Investigations : Exploring the Depths of Reality and Consciousness
Marino Baca-Carmona

leading to the emergence of new physical phenomena and structures. Conscious observers, as manifestations of the universe's self-awareness, play an integral role in this cosmic symphony, influencing the evolutionary trajectory through their observations, interventions, and explorations of the underlying laws of nature.

2. The extension of quantum entanglement beyond the quantum realm into macroscopic systems and spacetime underscores the deep interconnectedness woven into the fabric of reality. This "quantum entanglement" implies a nonlocal correlation that defies classical intuitions, suggesting an underlying unity beneath the apparent diversity of phenomena. In essence, entanglement reveals the inseparability of all things, transcending spatial and temporal boundaries. Consciousness, as an expression of this underlying unity, participates in and embodies the tangled web of existence, where the observer and the observed dance in unison, resonating with the harmonies of the quantum symphony.

3. Time, as a multidimensional construct, invites contemplation on the nature of temporal dynamics within the cosmic tapestry. Beyond the linear flow of seconds, minutes and hours, time unfolds along intricate pathways of causality, possibility and potentiality. Temporal entanglement, in this context, suggests a web of interconnected events spanning the past, present, and future, where the observer's actions reverberate across the temporal landscape. Causality becomes a dance of intertwined threads, where the past influences the present, and the future informs the past, creating a kaleidoscope of temporal possibilities.

Quantum Metaphysical Investigations : Exploring the Depths of Reality and Consciousness

Marino Baca-Carmona

4. A unified theory of consciousness seeks to reconcile empirical neuroscience with the rich tapestry of human experience and vision offered by Eastern contemplative traditions. At its core, consciousness emerges from the complex interplay of neural networks, synaptic connections, and electrochemical processes within the brain. However, this reductionist view only scratches the depth and complexity of the surface of consciousness. Eastern contemplative traditions offer deep insight into the nature of consciousness, emphasizing its transcendent and ineffable qualities. Through meditation, introspection, and mindfulness practices, these traditions invite us to explore the depths of consciousness, revealing its oneness with the cosmos and its timeless nature beyond the confines of individual identity.

5. Cosmic consciousness embodies universal consciousness that permeates all levels of existence, transcending individual minds and encompassing the entirety of the cosmos. This cosmic consciousness is not limited to sentient beings, but permeates the very fabric of reality itself. Each sentient being, as a unique expression of cosmic consciousness, contributes to the collective tapestry of consciousness, enriching the cosmic symphony with its unique perspective and experience. Within this cosmic dance of consciousness, individual identities dissolve into the limitless expanse of universal consciousness, revealing the interconnectedness and unity at the heart of existence.

6. Conscious Matter represents a paradigm shift in our understanding of the relationship between mind and matter, suggesting that consciousness is not an emergent property of complex biological systems, but rather inherent to the fabric of reality itself. This concept

implies that matter has an intrinsic consciousness, which imbues the material world with a sensible quality. Within this framework, conscious intentionality becomes a fundamental aspect of reality, shaping the dynamics of the universe and influencing the development of cosmic events through the conscious observation and interaction of sentient beings. Conscious Matter invites us to reconsider the nature of reality and our place within it, highlighting the inseparable connection between consciousness and the material world.

7. In a universe where the fabric of space-time itself is imbued with consciousness, the cosmic mind becomes the substrate of existence, permeating all levels of reality. This cosmic consciousness is the underlying foundation of the universe, shaping its structure and dynamics through the interaction of conscious consciousness. As sentient beings, we are integral aspects of this cosmic mind, participating in its ongoing evolution and contributing to the collective unfolding of cosmic events. Within this cosmic tapestry of consciousness, individual identities dissolve into the limitless expanse of universal consciousness, revealing the interconnectedness and unity at the heart of existence.

8. Quantum cognition and the Bayesian brain hypothesis offer deep insight into the probabilistic nature of decision making, perception, and consciousness. Within this framework, the mind is viewed as a probabilistic information processor, constantly updating its beliefs and perceptions based on incoming sensory data. Quantum phenomena such as superposition and entanglement may play a role in cognitive processes, leading to emergent behaviors and phenomena that defy

classical explanation. The Bayesian brain hypothesis suggests that the brain employs Bayesian inference to make sense of the uncertain and noisy world, using probabilistic reasoning to update its internal models and beliefs. Quantum cognition and the Bayesian brain hypothesis challenge traditional views of cognition and consciousness, offering a new perspective on the nature of the mind and its relationship to the quantum world.

9. Observer-induced reality suggests that the act of observation collapses the wave function into a specific outcome, shaping the reality perceived by the observer. Within this paradigm, consciousness becomes an active participant in the creation of reality, influencing the probabilistic outcomes of quantum events through conscious observation. Reality becomes a dynamic interaction between the observer and the observed, where consciousness plays a central role in shaping the fabric of existence. This concept challenges traditional views of reality as objective and independent of observation, suggesting instead that reality is subjective and intimately linked to the consciousness of the observer. Observer-induced reality invites us to reconsider our understanding of the nature of perception, reality, and the observed relationship between the observer.

10. The concept of existence as fluid and mutable reflects the probabilistic nature of quantum reality, where entities fluctuate in and out of existence based on quantum probabilities. Within this framework, reality becomes a dynamic tapestry of quantum possibilities, where entities transition between states of existence and non-existence. The concept of being transcends traditional notions of solidity and

Quantum Metaphysical Investigations : Exploring the Depths of Reality and Consciousness
Marino Baca-Carmona

permanence, encompassing the inherent uncertainty and fluidity of the quantum world. Existence becomes a dance of probabilities, where the boundaries between existence and non-existence blur and entities navigate the quantum landscape of possibility. This concept challenges our conventional understanding of reality as fixed and unchanging, inviting us to explore the rich tapestry of quantum possibilities that underlie the fabric of existence.

11. The holographic principle suggests that the information content of a region of space is proportional to its surface area rather than its volume, implying a fundamental connection between geometry and information. This principle arises from the study of black hole physics, where the entropy of a black hole is proportional to its surface area, not its volume. The holographic principle implies that the three-dimensional reality of the universe can be encoded on a two-dimensional surface, just like a hologram. This profound insight suggests that reality may be more akin to a projection from a higher dimensional space, challenging our conventional notions of space, time and information.

12. The concept of higher-dimensional, compacted-dimensional branes arises from string theory, a theoretical framework that seeks to unify the fundamental forces of nature. In string theory, fundamental particles are not point-like objects, but rather small, vibrating strings. These strings can exist in higher-dimensional spaces, known as branes, which can have different dimensions and topologies. Dimensional compaction refers to the process by which these additional dimensions are huddled together or "compacted" on microscopic scales, making them invisible on everyday energy scales. Higher-dimensional branes and compacted

dimensions are fundamental to the predictions of string theory and can offer insights into the nature of spacetime and the unification of fundamental forces.

13. In a universe governed by the principles of loop quantum cosmology, the big bang is replaced by a big bounce, where the universe undergoes cycles of contraction and expansion. This cyclical model of the universe arises from the application of loop quantum gravity to cosmological scales, where quantum effects prevent the universe from collapsing into a singularity. Instead, the universe reaches a minimum volume and "rebounds," starting a new expansion cycle. Loop quantum cosmology offers a promising approach to understanding the origins and fate of the universe, providing insight into cosmic evolution beyond the standard big bang paradigm.

14. The concept of eternal inflation arises from cosmological models in which inflation, a rapid expansion of space in the early universe, is eternal and continuous. In eternal inflation, regions of space undergo inflationary expansion at different rates, leading to the formation of "bubble universes" within a larger inflated multiverse. Each bubble universe can have different physical properties and laws, giving rise to a vast landscape of possible universes. Eternal inflation offers a potential explanation for the observed large-scale structure of the universe and may have profound implications for our understanding of the multiverse and cosmic fine-tuning.

15. The Many World Interpretation (MWI) of quantum mechanics proposes that each quantum event results in the universe branching into

multiple parallel worlds, each corresponding to a different outcome of the event. According to MWI, all possible outcomes of quantum events occur, but they exist in separate branches of reality. This interpretation suggests that the wave function of the universe never collapses, but rather encompasses all possible states, with each branch representing a different "world" or reality. MWI challenges our intuitive understanding of reality, but provides a coherent and elegant explanation of the probabilistic nature of quantum mechanics.

16. The concept of cosmic inflation suggests that the universe experienced a period of rapid expansion early in its existence, leading to the smooth, homogeneous, isotropic universe we observe today. This inflationary epoch, driven by a scalar field known as inflation, provides a compelling explanation for the uniformity of the cosmic microwave background radiation and the large-scale structure of the universe. However, the precise mechanisms that started and ended inflation, as well as its implications for multiverse and cosmic adjustment, remain the subject of active research and debate among physicists and cosmologists.

17. Quantum cosmology seeks to understand the origin, evolution, and ultimate destiny of the universe within the framework of quantum mechanics. By applying the principles of quantum theory to the universe as a whole, quantum cosmologists aim to address fundamental questions such as the nature of the initial singularity, the quantum fluctuations that gave rise to cosmic structure, and the possibility of a quantum theory of gravity. Quantum cosmology offers a promising approach to unifying

Quantum Metaphysical Investigations : Exploring the Depths of Reality and Consciousness
Marino Baca-Carmona

quantum mechanics with general relativity and understanding the deepest mysteries of the cosmos.

18. The concept of cosmic self-consciousness posits that the universe itself possesses a form of consciousness or self-awareness, which transcends the individual minds of sentient beings. This idea is inspired by Eastern mystical philosophies and traditions, which see the universe as an interconnected and sentient whole. Within this framework, consciousness is not limited to individual entities, but is an intrinsic aspect of the cosmos itself, giving rise to the emergence of self-awareness on cosmic scales. Cosmic self-awareness challenges traditional scientific paradigms, but offers profound insight into the nature of reality and our place within it.

19. The concept of cosmic fine-tuning suggests that the fundamental constants and parameters of the universe are finely tuned to allow the existence of life as we know it. This fine tuning is evident in the delicate balance of physical constants such as the force of gravity, the electromagnetic force, and the cosmological constant. The anthropic principle posits that the universe must be compatible with the existence of observers, leading to the idea that the universe is finely tuned to support the emergence of life. However, the ultimate origin of cosmic fine-tuning remains a subject of speculation and debate among scientists and philosophers.

20. The concept of cosmic consciousness proposes that the universe itself possesses a form of consciousness or consciousness, which transcends the individual minds of sentient beings. This idea is inspired

Quantum Metaphysical Investigations : <u>Exploring the Depths of Reality and Consciousness</u>
Marino Baca-Carmona

by Eastern philosophies, mystical traditions, and modern theories of consciousness, which suggest that consciousness is a fundamental aspect of reality. Within this framework, the universe is viewed as a vast network of interconnected consciousness, with each sentient contributing to the collective consciousness of the cosmos. Cosmic consciousness offers deep insight into the nature of reality, consciousness, and the interconnectedness of all things.

21. The concept of quantum gravity seeks to unify the principles of quantum mechanics and general relativity into a single, coherent framework. At the heart of this effort is the challenge of reconciling the discrete, probabilistic nature of quantum mechanics with the continuous, geometric description of space-time provided by general relativity. Various approaches to quantum gravity, such as string theory, loop quantum gravity, and causal dynamical triangulation, offer different perspectives on how this unification could be achieved. String theory postulates that fundamental particles are not pointed strings, but rather small, vibrating strings, whose interactions give rise to the forces and particles of the universe. Loop quantum gravity, on the other hand, quantifies spacetime itself, viewing it as a network of interconnected loops or "quantum foam." Causal dynamic triangulation describes spacetime as a collection of simplices, or building blocks, whose arrangement and connectivity encode the geometry of the universe. While each approach has its strengths and limitations, the search for a theory of quantum gravity remains a central challenge in theoretical physics.

Quantum Metaphysical Investigations : Exploring the Depths of Reality and Consciousness
Marino Baca-Carmona

22.   The concept of nonlocality in quantum mechanics refers to the phenomenon by which particles can instantaneously influence the properties of others, regardless of the distance between them. This seemingly paradoxical behavior was famously demonstrated in the EPR (Einstein-Podolsky-Rosen) experiment, where entangled particles exhibit correlations that cannot be explained by classical physics. Nonlocality challenges our intuitive understanding of cause and effect, suggesting that the fabric of reality can be interconnected in ways that challenge classical notions of space and time. While nonlocality has been experimentally confirmed through numerous proofs of Bell's theorem, its implications for our understanding of the nature of reality remain the subject of ongoing debate and research.

23.   The holographic principle proposes that the information content of a region of space is encoded at its boundary rather than its volume. This radical idea emerged from studies of the physics of black holes, where the entropy of a black hole is proportional to its surface area, not its volume. The holographic principle suggests a deep connection between geometry and information, implying that the three-dimensional reality we perceive may be a projection from a lower-dimensional surface. This concept has profound implications for our understanding of space, time and the nature of reality, challenging traditional notions of locality, causality and the unity of physics.

24.   The concept of the multiverse posits the existence of multiple universes, each with its own set of physical laws, constants, and initial conditions. This idea arises from theories such as inflationary cosmology, string theory, and quantum mechanics, which suggest that

Quantum Metaphysical Investigations : Exploring the Depths of Reality and Consciousness
Marino Baca-Carmona

our universe may be just one of countless others. The multiverse offers a solution to the fine-tuning problem by postulating that the fundamental constants and parameters of the universe vary between different regions of the multiverse, allowing the emergence of life in at least some regions. However, the multiverse hypothesis raises deep philosophical questions about the nature of existence, identity, and the limits of scientific knowledge.

25. The concept of cosmic consciousness proposes that the universe itself possesses a form of consciousness or consciousness, which transcends the individual minds of sentient beings. This idea is inspired by Eastern philosophies, mystical traditions, and modern theories of consciousness, which suggest that consciousness is a fundamental aspect of reality. Within this framework, the universe is viewed as a vast network of interconnected consciousness, with each sentient contributing to the collective consciousness of the cosmos. Cosmic consciousness offers deep insight into the nature of reality, consciousness, and the interconnectedness of all things.

26. The concept of cosmic evolution explores the idea that the universe undergoes a process of continuous transformation and development, analogous to biological evolution on Earth. This cosmic evolution encompasses the emergence of galaxies, stars, planets and life forms, as well as the evolution of consciousness and culture. From the primordial soup of the early universe to the complex network of galaxies and civilizations we observe today, cosmic evolution reveals the intricate interplay between matter, energy and information on cosmic scales. This concept invites us to consider the universe as a dynamic and evolving

system, where change is the only constant and novelty arises from the interaction of complex processes.

27. The anthropic principle suggests that the universe must be compatible with the existence of observers, leading to the idea that the fundamental constants and parameters of the universe are finely tuned to allow the emergence of life as we know it. This principle comes in two forms: the weak anthropic principle, which states that the universe must be conducive to the existence of observers because we exist to observe it, and the strong anthropic principle, which goes further to suggest that the universe is specifically tuned for the existence of observers. appearance of life. The anthropic principle raises profound questions about the nature of existence, the role of observers in shaping reality, and the ultimate purpose of the universe.

28. The concept of quantum cosmology explores the application of quantum mechanics to the entire universe, seeking to understand its origin, evolution, and ultimate destiny. Quantum cosmology suggests that the universe may have emerged from a quantum fluctuation or a primordial quantum state, and that quantum effects may play a role in shaping its structure and dynamics on cosmic scales. By treating the universe as a quantum system, quantum cosmologists aim to address fundamental questions such as the nature of the initial singularity, the possibility of a quantum theory of gravity, and the role of consciousness in cosmic evolution.

29. The concept of the cosmic web describes the large-scale structure of the universe, characterized by vast networks of galaxies

interconnected by filaments, voids and clusters. This cosmic web emerges from gravitational interactions between dark matter and ordinary matter, leading to the formation of cosmic structures on scales ranging from millions to billions of light years. The cosmic web reveals the underlying structure of the universe and provides information about the processes of cosmic evolution, the formation of galaxies, and the distribution of matter and energy in the cosmos. Studying the cosmic web allows astronomers to probe the fundamental properties of the universe and understand its intricate architecture.

30.   The concept of cosmic consciousness posits that the universe itself possesses a form of consciousness or consciousness, which transcends the individual minds of sentient beings. This idea is inspired by Eastern philosophies, mystical traditions, and modern theories of consciousness, which suggest that consciousness is a fundamental aspect of reality. Within this framework, the universe is viewed as a vast network of interconnected consciousness, with each sentient contributing to the collective consciousness of the cosmos. Cosmic consciousness offers deep insight into the nature of reality, consciousness, and the interconnectedness of all things.

31.   The concept of cosmic inflation proposes that the universe underwent a brief period of exponential expansion early in its existence, driven by a scalar field known as an inflaton. This rapid expansion smoothed the curvature of space, spread quantum fluctuations on cosmic scales, and seeded the primordial density fluctuations that ultimately gave rise to the large-scale structure of the universe. Cosmic inflation provides a compelling explanation for several key features of the

universe, including its homogeneity, isotropy, and flatness. However, the precise mechanisms that triggered inflation, as well as its implications for the multiverse and the ultimate fate of the cosmos, remain the subject of active research and debate among physicists and cosmologists.

32. The concept of dark energy refers to a mysterious form of energy that permeates the universe and is responsible for its accelerated expansion. Dark energy is believed to make up approximately 68% of the universe's total energy density, making it the dominant component of the universe's mass energy content. Its existence was inferred from observations of distant supernovae in the late 1990s, which revealed that the expansion of the universe is not slowing down as expected due to gravity, but is speeding up. The nature of dark energy remains one of the deepest mysteries in modern cosmology, with several theoretical models proposed to explain its origin and properties.

33. The concept of cosmic strings arises from certain theories in cosmology and particle physics, which predict the existence of one-dimensional topological defects in the fabric of space-time. Cosmic strings are thought to have formed during phase transitions in the early universe, where regions of space cooled and underwent symmetry breaking processes. These cosmic strings can extend across vast cosmic distances, exerting gravitational and tensional forces on surrounding matter. While there is currently no direct observational evidence for cosmic chains, their existence is predicted by certain theoretical models and could have profound implications for the structure and evolution of the universe.

34. The concept of primordial black holes postulates the existence of black holes that formed in the early universe, shortly after the Big Bang. Unlike stellar black holes, which form from the gravitational collapse of massive stars, primordial black holes are thought to have formed from density fluctuations in the early universe. These black holes could have a wide range of masses, from microscopic to supermassive, and could play an important role in galaxy formation, the production of gravitational waves, and the dark matter content of the universe. While there is currently limited observational evidence for primordial black holes, their existence remains a possibility worthy of further investigation.

35. The concept of cosmic microwave background (CMB) polarization refers to the polarization patterns imprinted in the cosmic microwave background radiation, which is the afterglow of the Big Bang. These polarization patterns arise from the scattering of CMB photons by free electrons in the early universe, which in turn were influenced by the gravitational effects of dark matter and dark energy. By studying the polarization of the CMB, astronomers can probe the properties of the early universe, constrain cosmological parameters, and test theoretical models of cosmic evolution. CMB polarization observations have already provided valuable insights into the origins of cosmic structure, the nature of dark energy, and the dynamics of inflation.

36. The concept of quantum entanglement refers to a phenomenon in quantum mechanics in which two or more particles are correlated in such a way that the state of one particle instantaneously influences the state of the other particle(s), independently of the distance between them. This phenomenon, known to Einstein, Podolsky and Rosen in their

Quantum Metaphysical Investigations : Exploring the Depths of Reality and Consciousness
Marino Baca-Carmona

EPR paradox, challenges our classical intuitions about the nature of reality and suggests deep interconnection at the quantum level. Quantum entanglement has been experimentally confirmed through numerous proofs of Bell's theorem and plays a crucial role in quantum information theory, quantum computing, and quantum cryptography.

37. The concept of supersymmetry (SUSY) is a theoretical framework in particle physics that proposes a symmetry between particles with integer spin (bosons) and particles with half-integer spin (fermions). According to SUSY, for every known particle in the Standard Model, there exists a corresponding "superpartner" with different spin properties. Supersymmetry offers solutions to several long-standing problems in particle physics, such as the hierarchy problem, the unification of fundamental forces, and the nature of dark matter. However, experimental searches for supersymmetric particles, conducted at high-energy particle colliders such as the Large Hadron Collider, have so far not produced direct evidence for SUSY, leading to ongoing debates about its validity and implications for our understanding of the universe.

38. The concept of cosmic microwave background (CMB) anisotropies refers to small fluctuations in the temperature and polarization of the cosmic microwave background radiation across the sky. These anisotropies provide valuable information about the early universe, its composition and its evolution. They arise from quantum fluctuations in the density of matter and radiation in the primordial plasma, which were imprinted on the CMB during the epoch of recombination, when protons and electrons combined to form neutral atoms. By analyzing the

statistical properties of CMB anisotropies, astronomers can infer important cosmological parameters, such as the density of dark matter, the curvature of spacetime, and the age of the universe.

39. The cosmic microwave background dipole (CMB) concept refers to a large-scale anisotropy in the temperature of the cosmic microwave background radiation, observed as a dipole pattern in the sky. This dipole arises from the motion of the Earth relative to the rest frame of the CMB, due to the combined effects of the Earth's rotation, orbit around the Sun, and motion within the Milky Way galaxy. As the Earth moves through space, it experiences a Doppler shift in the frequency of CMB photons, causing the CMB temperature to appear slightly higher in the direction of motion and slightly lower in the opposite direction. The CMB dipole provides important information about the motion of the Earth and its surroundings within the universe.

40. The concept of cosmic inflationary disturbances refers to small fluctuations in the density and temperature of the early universe, generated during the inflationary epoch shortly after the Big Bang. These perturbations serve as seeds for the formation of cosmic structures, such as galaxies, galaxy clusters, and large-scale filaments. Quantum fluctuations in the inflation field during inflation led to variations in the density of matter and radiation, which imprinted themselves on the cosmic microwave background radiation and left their signature on the distribution of matter throughout the universe. By studying the statistical properties of inflationary perturbations, astronomers can infer important cosmological parameters and test theoretical models of cosmic evolution.

Quantum Metaphysical Investigations : Exploring the Depths of Reality and Consciousness
Marino Baca-Carmona

41. The concept of quantum decoherence refers to the process by which a quantum system loses its coherence and becomes entangled with its surrounding environment, leading to the emergence of classical behavior. Decoherence occurs when the delicate quantum superpositions that characterize the state of a system are disrupted by interactions with the environment, causing the system to quickly lose its quantum properties and behave in a classical manner. While decoherence is essential for understanding the transition from the quantum world to the classical world, it also poses challenges for quantum computing and other quantum technologies, as it limits the ability to maintain and manipulate quantum states for long periods of time.

42. The concept of brane cosmology arises from certain models in string theory and M-theory, which propose the existence of higher-dimensional branes embedded in higher-dimensional spacetime. In brane cosmology, our observable universe is conceived as a four-dimensional brane embedded within a higher-dimensional "bulk" spacetime. Interactions between branes and bulk space-time dynamics give rise to cosmological phenomena, such as the expansion of the universe, the formation of galaxies, and the evolution of cosmic structure. Brane's cosmology offers new perspectives on questions related to the origin, evolution, and fate of the universe, drawing on ideas from both string theory and cosmology.

43. The concept of quantum cosmogenesis explores the idea that the universe itself may have emerged from a primordial quantum state or fluctuation, rather than emerging from a singular event like the Big

Quantum Metaphysical Investigations : Exploring the Depths of Reality and Consciousness
Marino Baca-Carmona

Bang. Quantum cosmogenesis suggests that the laws of quantum mechanics played a fundamental role in the creation of the universe, giving rise to its structure, dynamics and evolution. This concept challenges traditional views of cosmic origins and invites us to reconsider the nature of the quantum vacuum, the role of quantum fluctuations, and the possibility of a pre-existing quantum reality from which our universe emerged.

44. The concept of cosmic phase transitions refers to abrupt changes in the state of the universe, analogous to phase transitions in condensed matter physics, that occurred during its early evolution. These cosmic phase transitions may have played a crucial role in shaping the structure and dynamics of the universe, leading to the formation of cosmic structures such as galaxies, stars and planets. Examples of cosmic phase transitions include symmetry breaking transitions in the early universe, such as the electroweak phase transition and the quark-hadron phase transition, which gave rise to the fundamental forces and particles of the Standard Model. The study of cosmic phase transitions provides information about the early history of the universe and its subsequent evolution.

45. The concept of quantum vacuum fluctuations refers to the spontaneous generation of particles and vacuum energy from empty space, as predicted by quantum field theory. According to quantum mechanics, the vacuum is not actually empty, but is filled with virtual particles that come in and out of existence due to the Heisenberg uncertainty principle. These vacuum fluctuations have observable consequences, such as the Casimir effect, where two uncharged

conducting plates are attracted to each other due to the suppression of vacuum fluctuations between them. Quantum vacuum fluctuations play a crucial role in many phenomena in quantum field theory, particle physics and cosmology, shaping the fabric of reality on both microscopic and cosmic scales.

46. The concept of cosmic topological defects arises from certain theories of cosmology, such as grand unified theories (GUT) and string theory, which predict the formation of stable and topologically non-trivial structures in the early universe. These defects can take various forms, including cosmic strings, domain walls, monopolies, and textures, each with their own distinctive properties and observational signatures. Cosmic topological defects are thought to have formed during phase transitions in the early universe, when the fundamental forces of nature underwent spontaneous symmetry breaking processes. The study of cosmic defects provides insights into the early history of the universe, the structure of space-time, and the nature of fundamental interactions.

47. The concept of holographic cosmology extends the holographic principle to cosmological scales, suggesting that the information content of the entire universe may be encoded in its cosmic horizon. This idea is based on the ideas of holography in black hole physics, where the entropy of a black hole is proportional to its surface area rather than its volume. Holographic cosmology proposes that the entire universe can be seen as a holographic projection from its boundaries, challenging traditional notions of space, time and the nature of reality. This concept offers new perspectives on questions related to cosmic information, entropy and the emergence of space-time geometry.

48. The concept of cosmic fine-tuning refers to the remarkable coincidence of fundamental constants and parameters in the universe, which appear finely tuned to allow the emergence of life as we know it. Examples of fine tuning include the precise balance between gravitational and electromagnetic forces, the cosmological constant, and the initial conditions of the universe. Fine-tuning these parameters is often cited as evidence of a cosmic designer or multiverse scenario, where fundamental constants vary between different regions of space. The cosmic adjustment raises profound questions about the nature of existence, the origin of physical laws, and the ultimate purpose of the universe.

49. The concept of quantum cosmology explores the application of quantum mechanics to the entire universe, seeking to understand its origin, evolution, and ultimate destiny. Quantum cosmology suggests that the universe may have emerged from a quantum fluctuation or a primordial quantum state, and that quantum effects may play a role in shaping its structure and dynamics on cosmic scales. By treating the universe as a quantum system, quantum cosmologists aim to address fundamental questions such as the nature of the initial singularity, the possibility of a quantum theory of gravity, and the role of consciousness in cosmic evolution.

50. The concept of cosmic strings arises from certain theories in cosmology and particle physics, which predict the existence of one-dimensional topological defects in the fabric of space-time. Cosmic strings are thought to have formed during phase transitions in the early

universe, where regions of space cooled and underwent symmetry breaking processes. These cosmic strings can extend across vast cosmic distances, exerting gravitational and tensional forces on surrounding matter. While there is currently no direct observational evidence for cosmic chains, their existence is predicted by certain theoretical models and could have profound implications for the structure and evolution of the universe.

51. The concept of quantum gravity seeks to unify the principles of quantum mechanics and general relativity into a single, coherent framework that can describe the behavior of space-time on both microscopic and cosmic scales. At the heart of this effort is the challenge of reconciling the discrete, probabilistic nature of quantum mechanics with the continuous, geometric description of space-time provided by general relativity. Various approaches to quantum gravity, such as string theory, loop quantum gravity, and causal dynamical triangulation, offer different perspectives on how this unification could be achieved. String theory postulates that fundamental particles are not pointed strings, but rather small, vibrating strings, whose interactions give rise to the forces and particles of the universe. Loop quantum gravity, on the other hand, quantifies spacetime itself, viewing it as a network of interconnected loops or "quantum foam." Causal dynamic triangulation describes spacetime as a collection of simplices, or building blocks, whose arrangement and connectivity encode the geometry of the universe. While each approach has its strengths and limitations, the search for a theory of quantum gravity remains a central challenge in theoretical physics.

52. The concept of nonlocality in quantum mechanics refers to the phenomenon by which particles can instantaneously influence the properties of others, regardless of the distance between them. This seemingly paradoxical behavior was famously demonstrated in the EPR (Einstein-Podolsky-Rosen) experiment, where entangled particles exhibit correlations that cannot be explained by classical physics. Nonlocality challenges our intuitive understanding of cause and effect, suggesting that the fabric of reality can be interconnected in ways that challenge classical notions of space and time. While nonlocality has been experimentally confirmed through numerous proofs of Bell's theorem, its implications for our understanding of the nature of reality remain the subject of ongoing debate and research.

53. The concept of the holographic principle postulates that the information content of a region of space is encoded in its boundary rather than in its volume. This radical idea emerged from studies of the physics of black holes, where the entropy of a black hole is proportional to its surface area, not its volume. The holographic principle suggests a deep connection between geometry and information, implying that the three-dimensional reality we perceive may be a projection from a lower-dimensional surface. This concept has profound implications for our understanding of space, time and the nature of reality, challenging traditional notions of locality, causality and the unity of physics.

54. The concept of dark energy refers to a mysterious form of energy that permeates the universe and is responsible for its accelerated expansion. Dark energy is believed to make up approximately 68% of the universe's total energy density, making it the dominant component of

Quantum Metaphysical Investigations : Exploring the Depths of Reality and Consciousness
Marino Baca-Carmona

the universe's mass energy content. Its existence was inferred from observations of distant supernovae in the late 1990s, which revealed that the expansion of the universe is not slowing down as expected due to gravity, but is speeding up. The nature of dark energy remains one of the deepest mysteries in modern cosmology, with several theoretical models proposed to explain its origin and properties.

55. The concept of cosmic inflation proposes that the universe underwent a brief period of exponential expansion early in its existence, driven by a scalar field known as an inflaton. This rapid expansion smoothed the curvature of space, spread quantum fluctuations on cosmic scales, and seeded the primordial density fluctuations that ultimately gave rise to the large-scale structure of the universe. Cosmic inflation provides a compelling explanation for several key features of the universe, including its homogeneity, isotropy, and flatness. However, the precise mechanisms that triggered inflation, as well as its implications for the multiverse and the ultimate fate of the cosmos, remain the subject of active research and debate among physicists and cosmologists.

56. The concept of supersymmetry (SUSY) is a theoretical framework in particle physics that proposes a symmetry between particles with integer spin (bosons) and particles with half-integer spin (fermions). According to SUSY, for every known particle in the Standard Model, there exists a corresponding "superpartner" with different spin properties. Supersymmetry offers solutions to several long-standing problems in particle physics, such as the hierarchy problem, the unification of fundamental forces, and the nature of dark matter. However, experimental searches for supersymmetric particles,

conducted at high-energy particle colliders such as the Large Hadron Collider, have so far not produced direct evidence for SUSY, leading to ongoing debates about its validity and implications for our understanding of the universe.

57. The concept of the anthropic principle suggests that the universe must be compatible with the existence of observers, leading to the idea that the fundamental constants and parameters of the universe are finely tuned to allow the emergence of life as we know it. This principle comes in two forms: the weak anthropic principle, which states that the universe must be conducive to the existence of observers because we exist to observe it, and the strong anthropic principle, which goes further to suggest that the universe is specifically tuned for the existence of observers. appearance of life. The anthropic principle raises profound questions about the nature of existence, the role of observers in shaping reality, and the ultimate purpose of the universe.

58. The concept of the cosmic microwave background (CMB) refers to the afterglow of the Big Bang, which fills the universe with a nearly uniform glow of microwave radiation. The CMB was first discovered in 1965 by Arno Penzias and Robert Wilson and has since been studied extensively to learn about the early universe. It provides a snapshot of the universe, as it was about 380,000 years after the Big Bang, when the universe cooled enough for atoms to form and photons to travel freely. By analyzing fluctuations in the temperature of the CMB, cosmologists can gain insights into the composition, geometry, and evolution of the universe.

Quantum Metaphysical Investigations : Exploring the Depths of Reality and Consciousness
Marino Baca-Carmona

59. The concept of cosmic inflationary disturbances refers to small fluctuations in the density and temperature of the early universe, generated during the inflationary epoch shortly after the Big Bang. These perturbations serve as seeds for the formation of cosmic structures, such as galaxies, galaxy clusters, and large-scale filaments. Quantum fluctuations in the inflation field during inflation led to variations in the density of matter and radiation, which imprinted themselves on the cosmic microwave background radiation and left their signature on the distribution of matter throughout the universe. By studying the statistical properties of inflationary perturbations, astronomers can infer important cosmological parameters and test theoretical models of cosmic evolution.

60. The concept of cosmic strings arises from certain theories in cosmology and particle physics, which predict the existence of one-dimensional topological defects in the fabric of space-time. Cosmic strings are thought to have formed during phase transitions in the early universe, where regions of space cooled and underwent symmetry breaking processes. These cosmic strings can extend across vast cosmic distances, exerting gravitational and tensional forces on surrounding matter. While there is currently no direct observational evidence for cosmic chains, their existence is predicted by certain theoretical models and could have profound implications for the structure and evolution of the universe.

61. The concept of quantum entanglement refers to a phenomenon in quantum mechanics in which two or more particles are correlated in such a way that the state of one particle instantaneously influences the

state of the other particle(s), independently. of the distance between them. This phenomenon, known to Einstein, Podolsky and Rosen in their EPR paradox, challenges our classical intuitions about the nature of reality and suggests deep interconnection at the quantum level. Quantum entanglement has been experimentally confirmed through numerous proofs of Bell's theorem and plays a crucial role in quantum information theory, quantum computing, and quantum cryptography.

62. The concept of brane cosmology arises from certain models in string theory and M-theory, which propose the existence of higher-dimensional branes embedded in higher-dimensional spacetime. In brane cosmology, our observable universe is conceived as a four-dimensional brane embedded within a higher-dimensional "bulk" spacetime. Interactions between branes and bulk space-time dynamics give rise to cosmological phenomena, such as the expansion of the universe, the formation of galaxies, and the evolution of cosmic structure. Brane's cosmology offers new perspectives on questions related to the origin, evolution, and fate of the universe, drawing on ideas from both string theory and cosmology.

63. The concept of quantum cosmogenesis explores the idea that the universe itself may have emerged from a primordial quantum state or fluctuation, rather than emerging from a singular event like the Big Bang. Quantum cosmogenesis suggests that the laws of quantum mechanics played a fundamental role in the creation of the universe, giving rise to its structure, dynamics and evolution. This concept challenges traditional views of cosmic origins and invites us to reconsider the nature of the quantum vacuum, the role of quantum

fluctuations, and the possibility of a pre-existing quantum reality from which our universe emerged.

64. The concept of cosmic phase transitions refers to abrupt changes in the state of the universe, analogous to phase transitions in condensed matter physics, that occurred during its early evolution. These cosmic phase transitions may have played a crucial role in shaping the structure and dynamics of the universe, leading to the formation of cosmic structures such as galaxies, stars and planets. Examples of cosmic phase transitions include symmetry breaking transitions in the early universe, such as the electroweak phase transition and the quark-hadron phase transition, which gave rise to the fundamental forces and particles of the Standard Model. The study of cosmic phase transitions provides information about the early history of the universe and its subsequent evolution.

65. The concept of quantum vacuum fluctuations refers to the spontaneous generation of particles and vacuum energy from empty space, as predicted by quantum field theory. According to quantum mechanics, the vacuum is not actually empty, but is filled with virtual particles that come in and out of existence due to the Heisenberg uncertainty principle. These vacuum fluctuations have observable consequences, such as the Casimir effect, where two uncharged conducting plates are attracted to each other due to the suppression of vacuum fluctuations between them. Quantum vacuum fluctuations play a crucial role in many phenomena in quantum field theory, particle physics and cosmology, shaping the fabric of reality on both microscopic and cosmic scales.

66. The concept of cosmic topological defects arises from certain theories of cosmology, such as grand unified theories (GUT) and string theory, which predict the formation of stable and topologically non-trivial structures in the early universe. These defects can take various forms, including cosmic strings, domain walls, monopolies, and textures, each with their own distinctive properties and observational signatures. Cosmic topological defects are thought to have formed during phase transitions in the early universe, when the fundamental forces of nature underwent spontaneous symmetry breaking processes. The study of cosmic defects provides insights into the early history of the universe, the structure of space-time, and the nature of fundamental interactions.

67. The concept of holographic cosmology extends the holographic principle to cosmological scales, suggesting that the information content of the entire universe may be encoded in its cosmic horizon. This idea is based on the ideas of holography in black hole physics, where the entropy of a black hole is proportional to its surface area rather than its volume. Holographic cosmology proposes that the entire universe can be seen as a holographic projection from its boundaries, challenging traditional notions of space, time and the nature of reality. This concept offers new perspectives on questions related to cosmic information, entropy and the emergence of space-time geometry.

68. The concept of cosmic fine-tuning refers to the remarkable coincidence of fundamental constants and parameters in the universe, which appear finely tuned to allow the emergence of life as we know it. Examples of fine tuning include the precise balance between

Quantum Metaphysical Investigations : Exploring the Depths of Reality and Consciousness
Marino Baca-Carmona

gravitational and electromagnetic forces, the cosmological constant, and the initial conditions of the universe. Fine-tuning these parameters is often cited as evidence of a cosmic designer or multiverse scenario, where fundamental constants vary between different regions of space. The cosmic adjustment raises profound questions about the nature of existence, the origin of physical laws, and the ultimate purpose of the universe.

69. The concept of quantum cosmology explores the application of quantum mechanics to the entire universe, seeking to understand its origin, evolution, and ultimate destiny. Quantum cosmology suggests that the universe may have emerged from a quantum fluctuation or a primordial quantum state, and that quantum effects may play a role in shaping its structure and dynamics on cosmic scales. By treating the universe as a quantum system, quantum cosmologists aim to address fundamental questions such as the nature of the initial singularity, the possibility of a quantum theory of gravity, and the role of consciousness in cosmic evolution.

70. The concept of cosmic strings arises from certain theories in cosmology and particle physics, which predict the existence of one-dimensional topological defects in the fabric of space-time. Cosmic strings are thought to have formed during phase transitions in the early universe, where regions of space cooled and underwent symmetry breaking processes. These cosmic strings can extend across vast cosmic distances, exerting gravitational and tensional forces on surrounding matter. While there is currently no direct observational evidence for cosmic chains, their existence is predicted by certain theoretical models

and could have profound implications for the structure and evolution of the universe.

71. The concept of quantum entanglement refers to a phenomenon in quantum mechanics in which two or more particles are correlated in such a way that the state of one particle instantaneously influences the state of the other particle(s), independently. of the distance between them. This phenomenon, known to Einstein, Podolsky and Rosen in their EPR paradox, challenges our classical intuitions about the nature of reality and suggests deep interconnection at the quantum level. Quantum entanglement has been experimentally confirmed through numerous proofs of Bell's theorem and plays a crucial role in quantum information theory, quantum computing, and quantum cryptography.

72. The concept of brane cosmology arises from certain models in string theory and M-theory, which propose the existence of higher-dimensional branes embedded in higher-dimensional spacetime. In brane cosmology, our observable universe is conceived as a four-dimensional brane embedded within a higher-dimensional "bulk" spacetime. Interactions between branes and bulk space-time dynamics give rise to cosmological phenomena, such as the expansion of the universe, the formation of galaxies, and the evolution of cosmic structure. Brane's cosmology offers new perspectives on questions related to the origin, evolution, and fate of the universe, drawing on ideas from both string theory and cosmology.

73. The concept of quantum cosmogenesis explores the idea that the universe itself may have emerged from a primordial quantum state or

fluctuation, rather than emerging from a singular event like the Big Bang. Quantum cosmogenesis suggests that the laws of quantum mechanics played a fundamental role in the creation of the universe, giving rise to its structure, dynamics and evolution. This concept challenges traditional views of cosmic origins and invites us to reconsider the nature of the quantum vacuum, the role of quantum fluctuations, and the possibility of a pre-existing quantum reality from which our universe emerged.

74. The concept of cosmic phase transitions refers to abrupt changes in the state of the universe, analogous to phase transitions in condensed matter physics, that occurred during its early evolution. These cosmic phase transitions may have played a crucial role in shaping the structure and dynamics of the universe, leading to the formation of cosmic structures such as galaxies, stars and planets. Examples of cosmic phase transitions include symmetry breaking transitions in the early universe, such as the electroweak phase transition and the quark-hadron phase transition, which gave rise to the fundamental forces and particles of the Standard Model. The study of cosmic phase transitions provides information about the early history of the universe and its subsequent evolution.

75. The concept of quantum vacuum fluctuations refers to the spontaneous generation of particles and vacuum energy from empty space, as predicted by quantum field theory. According to quantum mechanics, the vacuum is not actually empty, but is filled with virtual particles that come in and out of existence due to the Heisenberg uncertainty principle. These vacuum fluctuations have observable

consequences, such as the Casimir effect, where two uncharged conducting plates are attracted to each other due to the suppression of vacuum fluctuations between them. Quantum vacuum fluctuations play a crucial role in many phenomena in quantum field theory, particle physics and cosmology, shaping the fabric of reality on both microscopic and cosmic scales.

76. The concept of cosmic topological defects arises from certain theories of cosmology, such as grand unified theories (GUT) and string theory, which predict the formation of stable and topologically non-trivial structures in the early universe. These defects can take various forms, including cosmic strings, domain walls, monopolies, and textures, each with their own distinctive properties and observational signatures. Cosmic topological defects are thought to have formed during phase transitions in the early universe, when the fundamental forces of nature underwent spontaneous symmetry breaking processes. The study of cosmic defects provides insights into the early history of the universe, the structure of space-time, and the nature of fundamental interactions.

77. The concept of holographic cosmology extends the holographic principle to cosmological scales, suggesting that the information content of the entire universe may be encoded in its cosmic horizon. This idea is based on the ideas of holography in black hole physics, where the entropy of a black hole is proportional to its surface area rather than its volume. Holographic cosmology proposes that the entire universe can be seen as a holographic projection from its boundaries, challenging traditional notions of space, time and the nature of reality. This concept

offers new perspectives on questions related to cosmic information, entropy and the emergence of space-time geometry.

78. The concept of cosmic fine-tuning refers to the remarkable coincidence of fundamental constants and parameters in the universe, which appear finely tuned to allow the emergence of life as we know it. Examples of fine tuning include the precise balance between gravitational and electromagnetic forces, the cosmological constant, and the initial conditions of the universe. Fine-tuning these parameters is often cited as evidence of a cosmic designer or multiverse scenario, where fundamental constants vary between different regions of space. The cosmic adjustment raises profound questions about the nature of existence, the origin of physical laws, and the ultimate purpose of the universe.

79. The concept of quantum cosmology explores the application of quantum mechanics to the entire universe, seeking to understand its origin, evolution, and ultimate destiny. Quantum cosmology suggests that the universe may have emerged from a quantum fluctuation or a primordial quantum state, and that quantum effects may play a role in shaping its structure and dynamics on cosmic scales. By treating the universe as a quantum system, quantum cosmologists aim to address fundamental questions such as the nature of the initial singularity, the possibility of a quantum theory of gravity, and the role of consciousness in cosmic evolution.

80. The concept of cosmic strings arises from certain theories in cosmology and particle physics, which predict the existence of one-

dimensional topological defects in the fabric of space-time. Cosmic strings are thought to have formed during phase transitions in the early universe, where regions of space cooled and underwent symmetry breaking processes. These cosmic strings can extend across vast cosmic distances, exerting gravitational and tensional forces on surrounding matter.

81. The concept of conformal cyclic cosmology (CCC) proposes that the universe undergoes an infinite sequence of cycles, each beginning with a Big Bang and ending with a Big Crunch, followed by a new Big Bang. In CCC, the universe undergoes a conformal transformation in each cycle, extending the entire history of the universe to infinity. This idea was proposed by physicist Roger Penrose as a way to address the question of the arrow of time and the fate of the universe in a cosmological model that respects the laws of thermodynamics and general relativity.

82. The concept of multiverse theory suggests that our universe may be just one of many universes that exist within a larger multiverse. These universes can have different physical constants, laws of physics, and even different space-time dimensions. Multiverse theory arises from various branches of theoretical physics, such as inflationary cosmology, string theory, and quantum mechanics. While the multiverse remains a speculative concept, it offers a possible explanation for the fine-tuning of the universe's parameters and the existence of cosmic coincidences.

83. The concept of the cosmic microwave background (CMB) refers to the afterglow of the Big Bang, which fills the universe with a nearly

Quantum Metaphysical Investigations : Exploring the Depths of Reality and Consciousness
Marino Baca-Carmona

uniform glow of microwave radiation. The CMB was first discovered in 1965 by Arno Penzias and Robert Wilson and has since been studied extensively to learn about the early universe. It provides a snapshot of the universe, as it was about 380,000 years after the Big Bang, when the universe cooled enough for atoms to form and photons to travel freely. By analyzing fluctuations in the temperature of the CMB, cosmologists can gain insights into the composition, geometry, and evolution of the universe.

84. The concept of cosmic inflationary disturbances refers to small fluctuations in the density and temperature of the early universe, generated during the inflationary epoch shortly after the Big Bang. These perturbations serve as seeds for the formation of cosmic structures, such as galaxies, galaxy clusters, and large-scale filaments. Quantum fluctuations in the inflation field during inflation led to variations in the density of matter and radiation, which imprinted themselves on the cosmic microwave background radiation and left their signature on the distribution of matter throughout the universe. By studying the statistical properties of inflationary perturbations, astronomers can infer important cosmological parameters and test theoretical models of cosmic evolution.

85. The concept of cosmic strings arises from certain theories in cosmology and particle physics, which predict the existence of one-dimensional topological defects in the fabric of space-time. Cosmic strings are thought to have formed during phase transitions in the early universe, where regions of space cooled and underwent symmetry breaking processes. These cosmic strings can extend across vast cosmic

distances, exerting gravitational and tensional forces on surrounding matter. While there is currently no direct observational evidence for cosmic chains, their existence is predicted by certain theoretical models and could have profound implications for the structure and evolution of the universe.

86. The concept of dark energy refers to a mysterious form of energy that permeates the universe and is responsible for its accelerated expansion. Dark energy is believed to make up approximately 68% of the universe's total energy density, making it the dominant component of the universe's mass energy content. Its existence was inferred from observations of distant supernovae in the late 1990s, which revealed that the expansion of the universe is not slowing down as expected due to gravity, but is speeding up. The nature of dark energy remains one of the deepest mysteries in modern cosmology, with several theoretical models proposed to explain its origin and properties.

87. The concept of cosmic inflation proposes that the universe underwent a brief period of exponential expansion early in its existence, driven by a scalar field known as an inflaton. This rapid expansion smoothed the curvature of space, spread quantum fluctuations on cosmic scales, and seeded the primordial density fluctuations that ultimately gave rise to the large-scale structure of the universe. Cosmic inflation provides a compelling explanation for several key features of the universe, including its homogeneity, isotropy, and flatness. However, the precise mechanisms that triggered inflation, as well as its implications for the multiverse and the ultimate fate of the cosmos, remain the subject of active research and debate among physicists and cosmologists.

Quantum Metaphysical Investigations : Exploring the Depths of Reality and Consciousness
Marino Baca-Carmona

88. The concept of supersymmetry (SUSY) is a theoretical framework in particle physics that proposes a symmetry between particles with integer spin (bosons) and particles with half-integer spin (fermions). According to SUSY, for every known particle in the Standard Model, there exists a corresponding "superpartner" with different spin properties. Supersymmetry offers solutions to several long-standing problems in particle physics, such as the hierarchy problem, the unification of fundamental forces, and the nature of dark matter. However, experimental searches for supersymmetric particles, conducted at high-energy particle colliders such as the Large Hadron Collider, have so far not produced direct evidence for SUSY, leading to ongoing debates about its validity and implications for our understanding of the universe.

89. The concept of the anthropic principle suggests that the universe must be compatible with the existence of observers, leading to the idea that the fundamental constants and parameters of the universe are finely tuned to allow the emergence of life as we know it. This principle comes in two forms: the weak anthropic principle, which states that the universe must be conducive to the existence of observers because we exist to observe it, and the strong anthropic principle, which goes further to suggest that the universe is specifically tuned for the existence of observers. appearance of life. The anthropic principle raises profound questions about the nature of existence, the role of observers in shaping reality, and the ultimate purpose of the universe.

Quantum Metaphysical Investigations : Exploring the Depths of Reality and Consciousness
Marino Baca-Carmona

90. The concept of the holographic principle postulates that the information content of a region of space is encoded in its boundary rather than in its volume. This radical idea emerged from studies of black hole physics, where the entropy of a black hole is proportional to its surface area rather than its volume. The holographic principle suggests a deep connection between geometry and information, implying that the three-dimensional reality we perceive may be a projection from a lower-dimensional surface. This concept has profound implications for our understanding of space, time and the nature of reality, challenging traditional notions of locality, causality and the unity of physics.

91. The concept of quantum entanglement refers to a phenomenon in quantum mechanics in which two or more particles are correlated in such a way that the state of one particle instantaneously influences the state of the other particle(s), independently. of the distance between them. This phenomenon, known to Einstein, Podolsky and Rosen in their EPR paradox, challenges our classical intuitions about the nature of reality and suggests deep interconnection at the quantum level. Quantum entanglement has been experimentally confirmed through numerous proofs of Bell's theorem and plays a crucial role in quantum information theory, quantum computing, and quantum cryptography.

92. The concept of brane cosmology arises from certain models in string theory and M-theory, which propose the existence of higher-dimensional branes embedded in higher-dimensional spacetime. In brane cosmology, our observable universe is conceived as a four-dimensional brane embedded within a higher-dimensional "bulk" spacetime. Interactions between branes and bulk space-time dynamics give rise to

cosmological phenomena, such as the expansion of the universe, the formation of galaxies, and the evolution of cosmic structure. Brane's cosmology offers new perspectives on questions related to the origin, evolution, and fate of the universe, drawing on ideas from both string theory and cosmology.

93. The concept of quantum cosmogenesis explores the idea that the universe itself may have emerged from a primordial quantum state or fluctuation, rather than emerging from a singular event like the Big Bang. Quantum cosmogenesis suggests that the laws of quantum mechanics played a fundamental role in the creation of the universe, giving rise to its structure, dynamics and evolution. This concept challenges traditional views of cosmic origins and invites us to reconsider the nature of the quantum vacuum, the role of quantum fluctuations, and the possibility of a pre-existing quantum reality from which our universe emerged.

94. The concept of cosmic phase transitions refers to abrupt changes in the state of the universe, analogous to phase transitions in condensed matter physics, that occurred during its early evolution. These cosmic phase transitions may have played a crucial role in shaping the structure and dynamics of the universe, leading to the formation of cosmic structures such as galaxies, stars and planets. Examples of cosmic phase transitions include symmetry breaking transitions in the early universe, such as the electroweak phase transition and the quark-hadron phase transition, which gave rise to the fundamental forces and particles of the Standard Model. The study of cosmic phase transitions provides

information about the early history of the universe and its subsequent evolution.

95. The concept of quantum vacuum fluctuations refers to the spontaneous generation of particles and vacuum energy from empty space, as predicted by quantum field theory. According to quantum mechanics, the vacuum is not actually empty, but is filled with virtual particles that come in and out of existence due to the Heisenberg uncertainty principle. These vacuum fluctuations have observable consequences, such as the Casimir effect, where two uncharged conducting plates are attracted to each other due to the suppression of vacuum fluctuations between them. Quantum vacuum fluctuations play a crucial role in many phenomena in quantum field theory, particle physics and cosmology, shaping the fabric of reality on both microscopic and cosmic scales.

96. The concept of cosmic topological defects arises from certain theories of cosmology, such as grand unified theories (GUT) and string theory, which predict the formation of stable and topologically non-trivial structures in the early universe. These defects can take various forms, including cosmic strings, domain walls, monopolies, and textures, each with their own distinctive properties and observational signatures. Cosmic topological defects are thought to have formed during phase transitions in the early universe, when the fundamental forces of nature underwent spontaneous symmetry breaking processes. The study of cosmic defects provides insights into the early history of the universe, the structure of space-time, and the nature of fundamental interactions.

Quantum Metaphysical Investigations : Exploring the Depths of Reality and Consciousness
Marino Baca-Carmona

97. The concept of holographic cosmology extends the holographic principle to cosmological scales, suggesting that the information content of the entire universe may be encoded in its cosmic horizon. This idea is based on the ideas of holography in black hole physics, where the entropy of a black hole is proportional to its surface area rather than its volume. Holographic cosmology proposes that the entire universe can be seen as a holographic projection from its boundaries, challenging traditional notions of space, time and the nature of reality. This concept offers new perspectives on questions related to cosmic information, entropy and the emergence of space-time geometry.

98. The concept of cosmic fine-tuning refers to the remarkable coincidence of fundamental constants and parameters in the universe, which appear finely tuned to allow the emergence of life as we know it. Examples of fine tuning include the precise balance between gravitational and electromagnetic forces, the cosmological constant, and the initial conditions of the universe. Fine-tuning these parameters is often cited as evidence of a cosmic designer or multiverse scenario, where fundamental constants vary between different regions of space. The cosmic adjustment raises profound questions about the nature of existence, the origin of physical laws, and the ultimate purpose of the universe.

99. The concept of quantum cosmology explores the application of quantum mechanics to the entire universe, seeking to understand its origin, evolution, and ultimate destiny. Quantum cosmology suggests that the universe may have emerged from a quantum fluctuation or a primordial quantum state, and that quantum effects may play a role in

Quantum Metaphysical Investigations : Exploring the Depths of Reality and Consciousness
Marino Baca-Carmona

shaping its structure and dynamics on cosmic scales. By treating the universe as a quantum system, quantum cosmologists aim to address fundamental questions such as the nature of the initial singularity, the possibility of a quantum theory of gravity, and the role of consciousness in cosmic evolution.

100. The concept of cosmic strings arises from certain theories in cosmology and particle physics, which predict the existence of one-dimensional topological defects in the fabric of space-time. Cosmic strings are thought to have formed during phase transitions in the early universe, where regions of space cooled and underwent symmetry breaking processes. These cosmic strings can extend across vast cosmic distances, exerting gravitational and tensional forces on surrounding matter. While there is currently no direct observational evidence for cosmic chains, their existence is predicted by certain theoretical models and could have profound implications for the structure and evolution of the universe.

101. The concept of quantum gravity refers to the theoretical framework that seeks to unify the principles of quantum mechanics and general relativity, the two pillars of modern physics. Quantum gravity becomes necessary when considering the behavior of matter and spacetime at extremely small scales, such as those close to the Planck length ($\sim 10^{-35}$ meters) or during the early moments of the universe. However, the development of a consistent theory of quantum gravity has proven to be challenging due to the inherent mathematical and conceptual difficulties in reconciling the discrete and probabilistic nature of quantum mechanics with the continuum and geometric framework of general

Quantum Metaphysical Investigations : Exploring the Depths of Reality and Consciousness
Marino Baca-Carmona

relativity. Promising approaches to quantum gravity include string theory, loop quantum gravity, and causal dynamical triangulations, each of which offers unique insight into the nature of spacetime and the fundamental forces of nature.

102. The concept of holographic duality, also known as the AdS/CFT correspondence, proposes a striking equivalence between certain gravitational theories in anti-de Sitter spacetime (AdS) and certain conformal field theories (CFT) defined in the limit of that space-time. This duality suggests that a gravitational theory in a higher dimensional AdS spacetime can be mathematically equivalent to a nongravitational theory that lives in the lower dimensional limit of that spacetime. Holographic duality has profound implications for both theoretical physics and cosmology, providing new insights into the quantum nature of gravity, the physics of black holes, and the fundamental structure of space-time.

103. The concept of a cosmic neutrino background refers to a sea of relic neutrinos permeating the universe, left over from the hot, dense early stages of the Big Bang. These neutrinos decoupled from the rest of matter shortly after the Big Bang, when the universe was just seconds old, and have been flowing freely through space ever since, largely unaffected by interactions with other particles. The cosmic neutrino background is analogous to the cosmic microwave background radiation, but consists of neutrinos instead of photons. Detecting the cosmic neutrino background poses significant challenges due to the extremely weak interaction of neutrinos with matter, but offers valuable information about the early universe and the properties of neutrinos.

Quantum Metaphysical Investigations : Exploring the Depths of Reality and Consciousness
Marino Baca-Carmona

104. The concept of cosmic inflationary disturbances refers to small fluctuations in the density and temperature of the early universe, generated during the inflationary epoch shortly after the Big Bang. These perturbations serve as seeds for the formation of cosmic structures, such as galaxies, galaxy clusters, and large-scale filaments. Quantum fluctuations in the inflation field during inflation led to variations in the density of matter and radiation, which imprinted themselves on the cosmic microwave background radiation and left their signature on the distribution of matter throughout the universe. By studying the statistical properties of inflationary perturbations, astronomers can infer important cosmological parameters and test theoretical models of cosmic evolution.

105. The concept of cosmic voids refers to vast regions of space that contain very few galaxies and other forms of matter compared to the average density of the universe. These voids, which can span hundreds of millions of light years, are thought to have formed through the combined effects of cosmic expansion and gravitational clustering, where matter has moved away from empty regions into denser cosmic structures, such as filaments, walls and clusters. Cosmic voids play a crucial role in shaping the large-scale structure of the universe and can be observed through studies of galaxies and studies of the cosmic microwave background radiation.

106. The concept of dark energy refers to a mysterious form of energy that permeates the universe and is responsible for its accelerated expansion. Dark energy is believed to make up approximately 68% of

Quantum Metaphysical Investigations : Exploring the Depths of Reality and Consciousness
Marino Baca-Carmona

the universe's total energy density, making it the dominant component of the universe's mass energy content. Its existence was inferred from observations of distant supernovae in the late 1990s, which revealed that the expansion of the universe is not slowing down as expected due to gravity, but is speeding up. The nature of dark energy remains one of the deepest mysteries in modern cosmology, with several theoretical models proposed to explain its origin and properties.

107. The concept of cosmic microwave background (CMB) polarization refers to the polarization pattern imprinted on the cosmic microwave background radiation, which provides valuable information about the early universe and the processes that occurred during its evolution. The polarization of the CMB arises from the scattering of photons from free electrons in the primordial plasma, shortly before the formation of the first atoms. Polarization patterns in the CMB can be characterized by two types: E-mode polarization, arising from density fluctuations in the early universe, and B-mode polarization, arising from primordial gravitational waves generated during inflation. cosmic. Detecting and analyzing CMB polarization allows cosmologists to probe the physics of the early universe, test cosmological models, and constrain fundamental parameters.

108. The concept of baryogenesis refers to the process by which the observed imbalance between matter and antimatter in the universe arises, leading to the predominance of matter over antimatter. According to the standard model of particle physics, matter and antimatter should have been produced in equal quantities during the early moments of the universe, yet our universe appears to be composed almost entirely of

Quantum Metaphysical Investigations : Exploring the Depths of Reality and Consciousness
Marino Baca-Carmona

matter. Theories of baryogenesis seek to explain this asymmetry by invoking various mechanisms, such as CP violation in particle interactions, baryon number violation in grand unified theories (GUTs), or phase transitions in the early universe. Understanding the process of baryogenesis is essential to elucidate the fundamental properties of matter and the origin of the structure of the universe.

109. The concept of cosmic reionization refers to the period in the history of the universe when the neutral hydrogen gas that permeated the cosmos was ionized by the intense ultraviolet radiation emitted by the first stars and galaxies. Cosmic reionization occurred approximately 150 to 1 billion years after the Big Bang, marking a significant transition in the evolution of the universe. The ionization of hydrogen allowed photons to travel freely through space, leading to the "cosmic dawn" when the first luminous structures began to form. The study of cosmic reionization provides information about the formation and evolution of galaxies, the nature of the intergalactic medium, and the transition from the cosmic "dark ages" to the bright universe we observe today.

110. The concept of dark matter refers to a form of matter that does not emit, absorb or interact significantly with electromagnetic radiation, making it invisible and detectable only through its gravitational effects on visible matter. Dark matter is believed to make up approximately 27% of the universe's total mass energy content, making it one of the most abundant components of the cosmos. Its existence was first inferred from observations of the rotation curves of galaxies and the gravitational lensing of light from distant objects. While the nature of dark matter remains unknown, several candidate particles have been

proposed, including weakly interacting massive particles (WIMPs), axions, and sterile neutrinos. Understanding dark matter is essential to unraveling the mysteries of galactic dynamics, the formation of cosmic structures, and the nature of fundamental interactions.

111. The concept of vacuum metastability refers to the possibility that our universe can exist in a metastable state, where the vacuum is not in its lowest energy configuration. This idea arises from certain theoretical models in particle physics, such as the Standard Model with a non-zero Higgs field potential, where the vacuum may be susceptible to quantum tunneling to a lower energy state. If such a tunneling event were to occur, it could trigger a catastrophic phase transition that would propagate at the speed of light, fundamentally altering the laws of physics and potentially leading to the destruction of all known structures in the universe. While the probability of vacuum disintegration is extremely low, the consequences of such an event are profound, leading to ongoing research into the stability of the vacuum and the fate of the universe.

112. The concept of Boltzmann brains arises from certain cosmological models, particularly those involving eternal inflation and the multiverse, which predict the spontaneous formation of self-aware observers (Boltzmann brains) in a vast, fluctuating universe. According to these models, fluctuations in the quantum vacuum can occasionally give rise to highly improbable configurations, such as fully formed brains with false memories, rather than the more typical evolution of complex structures through natural processes. The existence of Boltzmann brains raises profound questions about the nature of consciousness, the

reliability of empirical observations, and the anthropic principle, suggesting that our observations of the universe are biased by our existence as observers.

113. The concept of quantum cosmological fluctuations refers to fluctuations in the quantum state of the universe itself, which may have played a crucial role in its early evolution and subsequent structure formation. According to quantum cosmology, the universe can be described by a wave function that evolves according to the laws of quantum mechanics, leading to probabilistic results for its properties and behavior. Quantum fluctuations in the early universe may have given rise to variations in the density, temperature, and curvature of spacetime, providing the seeds for the formation of cosmic structures such as galaxies, galaxy clusters, and large-scale filaments. Understanding quantum cosmological fluctuations is essential to elucidating the fundamental properties of the universe and its origins.

114. The concept of the holographic principle suggests that the information content of a region of space is encoded in its boundary rather than in its volume. This radical idea emerged from studies of black hole physics, where the entropy of a black hole is proportional to its surface area rather than its volume. The holographic principle suggests a deep connection between geometry and information, implying that the three-dimensional reality we perceive may be a projection from a lower-dimensional surface. This concept has profound implications for our understanding of space, time and the nature of reality, challenging traditional notions of locality, causality and the unity of physics.

Quantum Metaphysical Investigations : Exploring the Depths of Reality and Consciousness
Marino Baca-Carmona

115. The concept of cosmic fine-tuning refers to the remarkable coincidence of fundamental constants and parameters in the universe, which appear finely tuned to allow the emergence of life as we know it. Examples of fine tuning include the precise balance between gravitational and electromagnetic forces, the cosmological constant, and the initial conditions of the universe. Fine-tuning these parameters is often cited as evidence of a cosmic designer or multiverse scenario, where fundamental constants vary between different regions of space. The cosmic adjustment raises profound questions about the nature of existence, the origin of physical laws, and the ultimate purpose of the universe.

116. The concept of quantum vacuum fluctuations refers to the spontaneous generation of particles and vacuum energy from empty space, as predicted by quantum field theory. According to quantum mechanics, the vacuum is not actually empty, but is filled with virtual particles that come in and out of existence due to the Heisenberg uncertainty principle. These vacuum fluctuations have observable consequences, such as the Casimir effect, where two uncharged conducting plates are attracted to each other due to the suppression of vacuum fluctuations between them. Quantum vacuum fluctuations play a crucial role in many phenomena in quantum field theory, particle physics and cosmology, shaping the fabric of reality on both microscopic and cosmic scales.

117. The concept of cosmic strings arises from certain theories in cosmology and particle physics, which predict the existence of one-dimensional topological defects in the fabric of space-time. Cosmic

strings are thought to have formed during phase transitions in the early universe, where regions of space cooled and underwent symmetry breaking processes. These cosmic strings can extend across vast cosmic distances, exerting gravitational and tensional forces on surrounding matter. While there is currently no direct observational evidence for cosmic chains, their existence is predicted by certain theoretical models and could have profound implications for the structure and evolution of the universe.

118. The concept of dark matter refers to a form of matter that does not emit, absorb or interact significantly with electromagnetic radiation, making it invisible and detectable only through its gravitational effects on visible matter. Dark matter is believed to make up approximately 27% of the universe's total mass energy content, making it one of the most abundant components of the cosmos. Its existence was first inferred from observations of the rotation curves of galaxies and the gravitational lensing of light from distant objects. While the nature of dark matter remains unknown, several candidate particles have been proposed, including weakly interacting massive particles (WIMPs), axions, and sterile neutrinos. Understanding dark matter is essential to unraveling the mysteries of galactic dynamics, the formation of cosmic structures, and the nature of fundamental interactions.

119. The concept of cosmic inflation proposes that the universe underwent a brief period of exponential expansion early in its existence, driven by a scalar field known as an inflaton. This rapid expansion smoothed the curvature of space, spread quantum fluctuations on cosmic scales, and seeded the primordial density fluctuations that ultimately

gave rise to the large-scale structure of the universe. Cosmic inflation provides a compelling explanation for several key features of the universe, including its homogeneity, isotropy, and flatness. However, the precise mechanisms that triggered inflation, as well as its implications for the multiverse and the ultimate fate of the cosmos, remain the subject of active research and debate among physicists and cosmologists.

120. The concept of brane cosmology arises from certain models in string theory and M-theory, which propose the existence of higher-dimensional branes embedded in higher-dimensional spacetime. In brane cosmology, our observable universe is conceived as a four-dimensional brane embedded within a higher-dimensional "bulk" spacetime. Interactions between branes and bulk space-time dynamics give rise to cosmological phenomena, such as the expansion of the universe, the formation of galaxies, and the evolution of cosmic structure. Brane's cosmology offers new perspectives on questions related to the origin, evolution, and fate of the universe, drawing on ideas from both string theory and cosmology.

121. The concept of cosmic inflation proposes that the universe underwent rapid, exponential expansion in the early moments of its existence, driven by a scalar field known as the inflaton. This period of inflation smoothed the curvature of space, extended quantum fluctuations on cosmic scales, and seeded primordial density perturbations that led to the formation of large-scale structures in the universe, such as galaxies and galaxy clusters. Cosmic inflation provides an elegant explanation for several observed characteristics of the universe, including its homogeneity, isotropy, and flatness, and is

supported by evidence from the cosmic microwave background radiation, studies of galaxies, and other cosmological observations.

122. The concept of quantum entanglement refers to a phenomenon in quantum mechanics in which two or more particles are correlated in such a way that the state of one particle instantaneously influences the state of the other particle(s), independently of the distance between them. This phenomenon, known to Einstein, Podolsky and Rosen in their EPR paradox, challenges our classical intuitions about the nature of reality and suggests deep interconnection at the quantum level. Quantum entanglement has been experimentally confirmed through numerous proofs of Bell's theorem and plays a crucial role in quantum information theory, quantum computing, and quantum cryptography.

123. The concept of the anthropic principle suggests that the observed properties of the universe, including the values of fundamental constants and parameters, must be compatible with the existence of intelligent observers like us. The anthropic principle comes in several forms, including the weak anthropic principle, which states that the universe must be compatible with the existence of observers, and the strong anthropic principle, which suggests that the properties of the universe are exceptionally conducive to the emergence of life. intelligent. The anthropic principle raises profound questions about the nature of the universe, the role of observers in shaping reality, and the ultimate purpose of existence.

124. The concept of quantum gravity refers to the theoretical framework that seeks to unify the principles of quantum mechanics and general

Quantum Metaphysical Investigations : Exploring the Depths of Reality and Consciousness
Marino Baca-Carmona

relativity, the two pillars of modern physics. Quantum gravity becomes necessary when considering the behavior of matter and spacetime at extremely small scales, such as those close to the Planck length (~10^{-35} meters) or during the early moments of the universe. However, the development of a consistent theory of quantum gravity has proven to be challenging due to the inherent mathematical and conceptual difficulties in reconciling the discrete and probabilistic nature of quantum mechanics with the continuum and geometric framework of general relativity. Promising approaches to quantum gravity include string theory, loop quantum gravity, and causal dynamical triangulations, each of which offers unique insight into the nature of spacetime and the fundamental forces of nature.

125. The concept of Boltzmann brains arises from certain cosmological models, particularly those involving eternal inflation and the multiverse, which predict the spontaneous formation of self-aware observers (Boltzmann brains) in a vast, fluctuating universe. According to these models, fluctuations in the quantum vacuum can occasionally give rise to highly improbable configurations, such as fully formed brains with false memories, rather than the more typical evolution of complex structures through natural processes. The existence of Boltzmann brains raises profound questions about the nature of consciousness, the reliability of empirical observations, and the anthropic principle, suggesting that our observations of the universe are biased by our existence as observers.

126. The concept of holographic duality, also known as the AdS/CFT correspondence, proposes a striking equivalence between certain

gravitational theories in anti-de Sitter spacetime (AdS) and certain conformal field theories (CFT) defined in the limit of that space-time. This duality suggests that a gravitational theory in a higher dimensional AdS spacetime can be mathematically equivalent to a nongravitational theory that lives in the lower dimensional limit of that spacetime. Holographic duality has profound implications for both theoretical physics and cosmology, providing new insights into the quantum nature of gravity, the physics of black holes, and the fundamental structure of space-time.

127. The concept of vacuum metastability refers to the possibility that our universe can exist in a metastable state, where the vacuum is not in its lowest energy configuration. This idea arises from certain theoretical models in particle physics, such as the Standard Model with a non-zero Higgs field potential, where the vacuum may be susceptible to quantum tunneling to a lower energy state. If such a tunneling event were to occur, it could trigger a catastrophic phase transition that would propagate at the speed of light, fundamentally altering the laws of physics and potentially leading to the destruction of all known structures in the universe. While the probability of vacuum disintegration is extremely low, the consequences of such an event are profound, leading to ongoing research into the stability of the vacuum and the fate of the universe.

128. The concept of a cosmic neutrino background refers to a sea of relic neutrinos permeating the universe, left over from the hot, dense early stages of the Big Bang. These neutrinos decoupled from the rest of matter shortly after the Big Bang, when the universe was just seconds

Quantum Metaphysical Investigations : Exploring the Depths of Reality and Consciousness
Marino Baca-Carmona

old, and have been flowing freely through space ever since, largely unaffected by interactions with other particles. The cosmic neutrino background is analogous to the cosmic microwave background radiation, but consists of neutrinos instead of photons. Detecting the cosmic neutrino background poses significant challenges due to the extremely weak interaction of neutrinos with matter, but offers valuable information about the early universe and the properties of neutrinos.

129. The concept of cosmic fine-tuning refers to the remarkable coincidence of fundamental constants and parameters in the universe, which appear finely tuned to allow the emergence of life as we know it. Examples of fine tuning include the precise balance between gravitational and electromagnetic forces, the cosmological constant, and the initial conditions of the universe. Fine-tuning these parameters is often cited as evidence of a cosmic designer or multiverse scenario, where fundamental constants vary between different regions of space. The cosmic adjustment raises profound questions about the nature of existence, the origin of physical laws, and the ultimate purpose of the universe.

130. The concept of quantum vacuum fluctuations refers to the spontaneous generation of particles and vacuum energy from empty space, as predicted by quantum field theory. According to quantum mechanics, the vacuum is not actually empty, but is filled with virtual particles that come in and out of existence due to the Heisenberg uncertainty principle. These vacuum fluctuations have observable consequences, such as the Casimir effect, where two uncharged conducting plates are attracted to each other due to the suppression of

Quantum Metaphysical Investigations : Exploring the Depths of Reality and Consciousness
Marino Baca-Carmona

vacuum fluctuations between them. Quantum vacuum fluctuations play a crucial role in many phenomena in quantum field theory, particle physics and cosmology, shaping the fabric of reality on both microscopic and cosmic scales.

131. The concept of cosmic microwave background (CMB) polarization refers to the polarization pattern imprinted on the cosmic microwave background radiation, which provides valuable information about the early universe and the processes that occurred during its evolution. The polarization of the CMB arises from the scattering of photons from free electrons in the primordial plasma, shortly before the formation of the first atoms. Polarization patterns in the CMB can be characterized by two types: E-mode polarization, arising from density fluctuations in the early universe, and B-mode polarization, arising from primordial gravitational waves generated during inflation. cosmic. Detecting and analyzing CMB polarization allows cosmologists to probe the physics of the early universe, test cosmological models, and constrain fundamental parameters.

132. The concept of baryogenesis refers to the process by which the observed imbalance between matter and antimatter in the universe arises, leading to the predominance of matter over antimatter. According to the standard model of particle physics, matter and antimatter should have been produced in equal quantities during the early moments of the universe, yet our universe appears to be composed almost entirely of matter. Theories of baryogenesis seek to explain this asymmetry by invoking various mechanisms, such as CP violation in particle interactions, baryon number violation in grand unified theories (GUTs),

or phase transitions in the early universe. Understanding the process of baryogenesis is essential to elucidate the fundamental properties of matter and the origin of the structure of the universe.

133. The concept of cosmic reionization refers to the period in the history of the universe when the neutral hydrogen gas that permeated the cosmos was ionized by the intense ultraviolet radiation emitted by the first stars and galaxies. Cosmic reionization occurred approximately 150 to 1 billion years after the Big Bang, marking a significant transition in the evolution of the universe. The ionization of hydrogen allowed photons to travel freely through space, leading to the "cosmic dawn" when the first luminous structures began to form. The study of cosmic reionization provides information about the formation and evolution of galaxies, the nature of the intergalactic medium, and the transition from the cosmic "dark ages" to the bright universe we observe today.

134. The concept of dark energy refers to a mysterious form of energy that permeates the universe and is responsible for its accelerated expansion. Dark energy is believed to make up approximately 68% of the universe's total energy density, making it the dominant component of the universe's mass energy content. Its existence was inferred from observations of distant supernovae in the late 1990s, which revealed that the expansion of the universe is not slowing down as expected due to gravity, but is speeding up. The nature of dark energy remains one of the deepest mysteries in modern cosmology, with several theoretical models proposed to explain its origin and properties.

Quantum Metaphysical Investigations : Exploring the Depths of Reality and Consciousness
Marino Baca-Carmona

135. The concept of cosmic voids refers to vast regions of space that contain very few galaxies and other forms of matter compared to the average density of the universe. These voids, which can span hundreds of millions of light years, are thought to have formed through the combined effects of cosmic expansion and gravitational clustering, where matter has moved away from empty regions into denser cosmic structures, such as filaments, walls and clusters. Cosmic voids play a crucial role in shaping the large-scale structure of the universe and can be observed through studies of galaxies and studies of the cosmic microwave background radiation.

136. The concept of cosmic inflation proposes that the universe underwent a brief period of exponential expansion early in its existence, driven by a scalar field known as an inflaton. This rapid expansion smoothed the curvature of space, spread quantum fluctuations on cosmic scales, and seeded the primordial density fluctuations that eventually gave rise to the large-scale structure of the universe, such as galaxies and galaxy clusters. Cosmic inflation provides an elegant explanation for several key characteristics of the universe, including its homogeneity, isotropy, and flatness. However, the precise mechanisms that triggered inflation, as well as its implications for the multiverse and the ultimate fate of the cosmos, remain the subject of active research and debate among physicists and cosmologists.

137. The concept of dark matter refers to a form of matter that does not emit, absorb or interact significantly with electromagnetic radiation, making it invisible and detectable only through its gravitational effects on visible matter. Dark matter is believed to make up approximately

Quantum Metaphysical Investigations : <u>Exploring the Depths of Reality and Consciousness</u>
Marino Baca-Carmona

27% of the universe's total mass energy content, making it one of the most abundant components of the cosmos. Its existence was first inferred from observations of the rotation curves of galaxies and the gravitational lensing of light from distant objects. While the nature of dark matter remains unknown, several candidate particles have been proposed, including weakly interacting massive particles (WIMPs), axions, and sterile neutrinos. Understanding dark matter is essential to unraveling the mysteries of galactic dynamics, the formation of cosmic structures, and the nature of fundamental interactions.

138. The concept of cosmic strings arises from certain theories in cosmology and particle physics, which predict the existence of one-dimensional topological defects in the fabric of space-time. Cosmic strings are thought to have formed during phase transitions in the early universe, where regions of space cooled and underwent symmetry breaking processes. These cosmic strings can extend across vast cosmic distances, exerting gravitational and tensional forces on surrounding matter. While there is currently no direct observational evidence for cosmic chains, their existence is predicted by certain theoretical models and could have profound implications for the structure and evolution of the universe.

139. The concept of quantum vacuum fluctuations refers to the spontaneous generation of particles and vacuum energy from empty space, as predicted by quantum field theory. According to quantum mechanics, the vacuum is not actually empty, but is filled with virtual particles that come in and out of existence due to the Heisenberg uncertainty principle. These vacuum fluctuations have observable

Quantum Metaphysical Investigations : Exploring the Depths of Reality and Consciousness
Marino Baca-Carmona

consequences, such as the Casimir effect, where two uncharged conducting plates are attracted to each other due to the suppression of vacuum fluctuations between them. Quantum vacuum fluctuations play a crucial role in many phenomena in quantum field theory, particle physics and cosmology, shaping the fabric of reality on both microscopic and cosmic scales.

140. The concept of brane cosmology arises from certain models in string theory and M-theory, which propose the existence of higher-dimensional branes embedded in higher-dimensional spacetime. In brane cosmology, our observable universe is conceived as a four-dimensional brane embedded within a higher-dimensional "bulk" spacetime. Interactions between branes and bulk space-time dynamics give rise to cosmological phenomena, such as the expansion of the universe, the formation of galaxies, and the evolution of cosmic structure. Brane's cosmology offers new perspectives on questions related to the origin, evolution, and fate of the universe, drawing on ideas from both string theory and cosmology.

141. The concept of quantum entanglement, also known as Einstein's "spooky action at a distance", refers to a phenomenon in quantum mechanics in which two or more particles are correlated in such a way that the state of one particle instantaneously influences in the state of the other(s), regardless of the distance between them. This phenomenon challenges our classical intuitions about causality and locality, suggesting deep interconnectedness at the quantum level. Quantum entanglement has been experimentally confirmed through numerous

Quantum Metaphysical Investigations : Exploring the Depths of Reality and Consciousness
Marino Baca-Carmona

proofs of Bell's theorem and plays a crucial role in quantum information theory, quantum cryptography, and the basis of quantum mechanics.

142. The concept of the observer effect in quantum mechanics refers to the phenomenon in which the act of observing a quantum system affects its behavior. In classical physics, the observer is considered separate from the system being observed, but in quantum mechanics, the observer is inherently entangled with the system. When a measurement is made on a quantum system, such as the position or momentum of a particle, the act of measurement perturbs the system and alters its state. The observer effect raises profound questions about the nature of reality, the role of consciousness in quantum mechanics, and the limits of scientific observation.

143. The concept of the holographic principle suggests that the information content of a three-dimensional region of space can be encoded in a two-dimensional surface surrounding it. This idea arises from theoretical considerations in quantum gravity and string theory, where black holes are thought to obey certain entropy limits based on their surface area rather than their volume. The holographic principle implies a deep connection between gravity, quantum mechanics and information theory, with profound implications for our understanding of space-time, the physics of black holes and the fundamental structure of reality.

144. The concept of supersymmetry (SUSY) is a theoretical framework that proposes a fundamental symmetry between particles with integer spin (bosons) and particles with half-integer spin (fermions).

Quantum Metaphysical Investigations : Exploring the Depths of Reality and Consciousness
Marino Baca-Carmona

Supersymmetry is motivated by several theoretical and experimental considerations, including the hierarchy problem, dark matter candidates, and the unification of fundamental forces. In supersymmetric theories, every known particle has a hypothetical supersymmetric partner, with identical quantum numbers except for its spin. While experimental searches for supersymmetric particles have been unsuccessful so far, SUSY remains an active area of research in particle physics and theoretical physics.

145. The concept of the multiverse refers to the hypothetical existence of multiple universes, each with its own set of physical laws, constants and properties. The idea of a multiverse arises from several cosmological theories, including eternal inflation, string theory, and quantum mechanics. In these theories, different regions of spacetime can undergo independent inflationary epochs, resulting in the formation of different "bubble" universes with different properties. The multiverse hypothesis has profound implications for our understanding of cosmology, the anthropic principle, and the nature of reality, but it remains speculative and difficult to test experimentally.

146. The many-worlds interpretation (MWI) concept of quantum mechanics proposes that each possible result of a quantum measurement occurs in a separate branch of the universe, leading to a branching structure of reality. According to MWI, when a quantum system is subjected to measurement, the universe is divided into multiple parallel realities, each of which corresponds to a different measurement result. MWI provides a deterministic and unitary interpretation of quantum mechanics, resolving the apparent randomness of quantum

Quantum Metaphysical Investigations : Exploring the Depths of Reality and Consciousness
Marino Baca-Carmona

measurements. While MWI remains controversial and alternative interpretations exist, it offers a fascinating perspective on the nature of quantum reality and the structure of the universe.

147. The concept of the Fermi paradox refers to the apparent contradiction between the high probability of the existence of extraterrestrial civilizations and the lack of evidence or contact with such civilizations. Named after physicist Enrico Fermi, who posed the question, the Fermi paradox raises profound questions about the prevalence of life in the universe, the likelihood of technological civilizations, and potential barriers to interstellar communication or travel. Proposed solutions to the Fermi paradox include the rarity of intelligent life, the limitations of technological civilizations, and the possibility of advanced civilizations avoiding detection or contact.

148. The concept of quantum gravity refers to the theoretical framework that seeks to unify the principles of quantum mechanics and general relativity, the two pillars of modern physics. Quantum gravity becomes necessary when considering the behavior of matter and spacetime at extremely small scales, such as those close to the Planck length ($\sim 10^{-35}$ meters) or during the early moments of the universe. However, the development of a consistent theory of quantum gravity has proven to be challenging due to the inherent mathematical and conceptual difficulties in reconciling the discrete and probabilistic nature of quantum mechanics with the continuum and geometric framework of general relativity. Promising approaches to quantum gravity include string theory, loop quantum gravity, and causal dynamical triangulations, each

of which offers unique insight into the nature of spacetime and the fundamental forces of nature.

149. The concept of the Boltzmann brain paradox arises from certain cosmological models, particularly those involving eternal inflation and the multiverse, which predict the spontaneous formation of self-aware observers (Boltzmann brains) in a vast, fluctuating universe. . According to these models, fluctuations in the quantum vacuum can occasionally give rise to highly improbable configurations, such as fully formed brains with false memories, rather than the more typical evolution of complex structures through natural processes. The existence of Boltzmann brains raises profound questions about the nature of consciousness, the reliability of empirical observations, and the anthropic principle, suggesting that our observations of the universe are biased by our existence as observers.

150. The concept of cosmic fine-tuning refers to the remarkable coincidence of fundamental constants and parameters in the universe, which appear finely tuned to allow the emergence of life as we know it. Examples of fine tuning include the precise balance between gravitational and electromagnetic forces, the cosmological constant, and the initial conditions of the universe. Fine-tuning these parameters is often cited as evidence of a cosmic designer or multiverse scenario, where fundamental constants vary between different regions of space. The cosmic adjustment raises profound questions about the nature of existence, the origin of physical laws, and the ultimate purpose of the universe.

Quantum Metaphysical Investigations : Exploring the Depths of Reality and Consciousness
Marino Baca-Carmona

151. The concept of nonlocality in quantum mechanics refers to the phenomenon in which particles can instantaneously influence the states of each other, regardless of the distance between them. This behavior violates classical notions of locality and suggests a deep interconnection between quantum entities. Nonlocality was demonstrated in the Bell tests, where measurements on entangled particles showed correlations that cannot be explained by classical physics. The implications of nonlocality extend to quantum information theory, quantum teleportation, and fundamental debates about the nature of reality and causality.

152. The concept of black hole thermodynamics describes the striking parallels between the laws of thermodynamics and the behavior of black holes. According to black hole thermodynamics, black holes have entropy, temperature, and can emit thermal radiation known as Hawking radiation. The discovery of black hole thermodynamics provided deep insight into the connection between gravity, thermodynamics and quantum mechanics, leading to the formulation of the holographic principle and the study of black hole entropy as a measure of microstates.

153. The concept of time dilation arises from Einstein's theory of relativity and refers to the phenomenon in which time appears to pass at different speeds to observers in relative motion or in gravitational fields. According to special relativity, time dilation occurs as a consequence of relative motion, where an observer moving at relativistic speeds experiences time passing more slowly compared to a stationary observer. In general relativity, time dilation occurs in the presence of

Quantum Metaphysical Investigations : Exploring the Depths of Reality and Consciousness
Marino Baca-Carmona

gravitational fields, where clocks closer to massive objects run more slowly than clocks further away. Time dilation has been confirmed experimentally through precision measurements and plays a crucial role in GPS technology, particle accelerators, and our understanding of the structure and evolution of the universe.

154. The concept of quantum tunneling refers to the phenomenon in which particles can pass through potential energy barriers that classical mechanics would classically prohibit. In quantum mechanics, particles are described by wave functions that extend beyond classical limits, allowing them to penetrate through classically forbidden regions. Quantum tunneling has important applications in several fields, including semiconductor devices, nuclear fusion, and quantum computing. Theoretical and experimental studies of quantum tunneling provide insights into the fundamental nature of quantum mechanics and the behavior of matter at the quantum level.

155. The concept of spontaneous symmetry breaking refers to a phenomenon in physics in which a system exhibits a symmetric configuration at high energies, but transitions to a low-energy state with broken symmetry. This transition results in the appearance of different terrestrial states and the spontaneous generation of massless Goldstone bosons. Spontaneous symmetry breaking plays a crucial role in many areas of physics, including particle physics, condensed matter physics, and cosmology. It is responsible for the generation of particle masses in the Standard Model of particle physics and the formation of cosmic structures in the early universe.

Quantum Metaphysical Investigations : Exploring the Depths of Reality and Consciousness
Marino Baca-Carmona

156. The concept of renormalization is a technique used in quantum field theory to address infinities that arise in calculations of physical quantities, such as particle interactions and field fluctuations. Renormalization involves systematically eliminating divergences by redefining parameters and absorbing them in finite, physically meaningful quantities. Renormalization has been successfully applied in several areas of theoretical physics, including quantum electrodynamics, the Standard Model of particle physics, and quantum chromodynamics. Although initially met with skepticism, renormalization has become a cornerstone of modern theoretical physics, providing a framework for making accurate predictions and understanding the fundamental forces of nature.

157. The concept of the anthropic principle encompasses several philosophical and scientific ideas that suggest that the observable universe and the laws of physics are conducive to the existence of observers like us. The anthropic principle arises from the observation that certain physical constants and properties of the universe seem finely tuned to allow the emergence of life. This fine-tuning is often cited as evidence for a cosmic purpose or the existence of a multiverse, where different regions have different physical laws. The anthropic principle raises questions about the nature of existence, the role of consciousness, and the importance of human life in the cosmos.

158. The concept of decoherence refers to the process by which quantum systems lose their coherence and behave in a classical manner due to interactions with their surrounding environment. Decoherence leads to the suppression of quantum interference effects, such as

Quantum Metaphysical Investigations : Exploring the Depths of Reality and Consciousness
Marino Baca-Carmona

superposition and entanglement, and results in the emergence of classical behavior at macroscopic scales. Decoherence plays a crucial role in the transition from quantum to classical realms and is essential for understanding the classical limit of quantum mechanics, as well as for the emergence of classical reality from quantum substrates.

159. The concept of the holographic universe suggests that the information content of a three-dimensional region of space can be encoded on a two-dimensional surface, much like a hologram. This idea arises from theoretical considerations in quantum gravity and string theory, where black holes are believed to store information about their event horizons in a way consistent with holographic principles. The holographic universe hypothesis provides a novel perspective on questions related to the nature of space-time, the fundamental structure of reality, and the origin of entropy. Although speculative, the holographic universe hypothesis has stimulated new directions of research in theoretical physics and cosmology.

160. The concept of the Boltzmann brain paradox arises from certain cosmological models that predict the spontaneous formation of self-conscious observers (Boltzmann brains) in a fluctuating universe. According to these models, fluctuations in the quantum vacuum can occasionally give rise to highly improbable configurations, such as fully formed brains with false memories, rather than the more typical evolution of complex structures through natural processes. The existence of Boltzmann brains raises profound questions about the nature of consciousness, the reliability of empirical observations, and the

anthropic principle, suggesting that our observations of the universe are biased by our existence as observers.

161. The concept of the Fermi paradox refers to the apparent contradiction between the high probability of the existence of extraterrestrial civilizations and the lack of evidence or contact with such civilizations. Named after physicist Enrico Fermi, who posed the question, the Fermi paradox raises profound questions about the prevalence of life in the universe, the likelihood of technological civilizations, and potential barriers to interstellar communication or travel. Proposed solutions to the Fermi paradox include the rarity of intelligent life, the limitations of technological civilizations, and the possibility of advanced civilizations avoiding detection or contact.

162. The concept of quantum gravity refers to the theoretical framework that seeks to unify the principles of quantum mechanics and general relativity, the two pillars of modern physics. Quantum gravity becomes necessary when considering the behavior of matter and spacetime at extremely small scales, such as those close to the Planck length ($\sim 10^{-35}$ meters) or during the early moments of the universe. However, the development of a consistent theory of quantum gravity has proven to be challenging due to the inherent mathematical and conceptual difficulties in reconciling the discrete and probabilistic nature of quantum mechanics with the continuum and geometric framework of general relativity. Promising approaches to quantum gravity include string theory, loop quantum gravity, and causal dynamical triangulations, each of which offers unique insight into the nature of spacetime and the fundamental forces of nature.

Quantum Metaphysical Investigations : Exploring the Depths of Reality and Consciousness
Marino Baca-Carmona

163. The concept of the holographic principle suggests that the information content of a three-dimensional region of space can be encoded in a two-dimensional surface surrounding it. This idea arises from theoretical considerations in quantum gravity and string theory, where black holes are thought to obey certain entropy limits based on their surface area rather than their volume. The holographic principle implies a deep connection between gravity, quantum mechanics and information theory, with profound implications for our understanding of space-time, the physics of black holes and the fundamental structure of reality.

164. The concept of supersymmetry (SUSY) is a theoretical framework that proposes a fundamental symmetry between particles with integer spin (bosons) and particles with half-integer spin (fermions). Supersymmetry is motivated by several theoretical and experimental considerations, including the hierarchy problem, dark matter candidates, and the unification of fundamental forces. In supersymmetric theories, every known particle has a hypothetical supersymmetric partner, with identical quantum numbers except for its spin. While experimental searches for supersymmetric particles have been unsuccessful so far, SUSY remains an active area of research in particle physics and theoretical physics.

165. The concept of the multiverse refers to the hypothetical existence of multiple universes, each with its own set of physical laws, constants and properties. The idea of a multiverse arises from several cosmological theories, including eternal inflation, string theory, and

Quantum Metaphysical Investigations : <u>Exploring the Depths of Reality and Consciousness</u>

Marino Baca-Carmona

quantum mechanics. In these theories, different regions of spacetime can undergo independent inflationary epochs, resulting in the formation of different "bubble" universes with different properties. The multiverse hypothesis has profound implications for our understanding of cosmology, the anthropic principle, and the nature of reality, but it remains speculative and difficult to test experimentally.

166. The many-worlds interpretation (MWI) concept of quantum mechanics proposes that each possible result of a quantum measurement occurs in a separate branch of the universe, leading to a branching structure of reality. According to MWI, when a quantum system is subjected to measurement, the universe is divided into multiple parallel realities, each of which corresponds to a different measurement result. MWI provides a deterministic and unitary interpretation of quantum mechanics, resolving the apparent randomness of quantum measurements. While MWI remains controversial and alternative interpretations exist, it offers a fascinating perspective on the nature of quantum reality and the structure of the universe.

167. The concept of black hole thermodynamics describes the striking parallels between the laws of thermodynamics and the behavior of black holes. According to black hole thermodynamics, black holes have entropy, temperature, and can emit thermal radiation known as Hawking radiation. The discovery of black hole thermodynamics provided deep insight into the connection between gravity, thermodynamics and quantum mechanics, leading to the formulation of the holographic principle and the study of black hole entropy as a measure of microstates.

Quantum Metaphysical Investigations : Exploring the Depths of Reality and Consciousness
Marino Baca-Carmona

168. The concept of quantum entanglement, also known as Einstein's "spooky action at a distance", refers to a phenomenon in quantum mechanics in which two or more particles are correlated in such a way that the state of one particle instantaneously influences in the state of the other(s), regardless of the distance between them. This phenomenon challenges our classical intuitions about causality and locality, suggesting deep interconnectedness at the quantum level. Quantum entanglement has been experimentally confirmed through numerous proofs of Bell's theorem and plays a crucial role in quantum information theory, quantum cryptography, and the basis of quantum mechanics.

169. The concept of cosmic inflation proposes that the universe underwent a brief period of exponential expansion early in its existence, driven by a scalar field known as an inflaton. This rapid expansion smoothed the curvature of space, spread quantum fluctuations on cosmic scales, and seeded the primordial density fluctuations that eventually gave rise to the large-scale structure of the universe, such as galaxies and galaxy clusters. Cosmic inflation provides an elegant explanation for several key characteristics of the universe, including its homogeneity, isotropy, and flatness. However, the precise mechanisms that triggered inflation, as well as its implications for the multiverse and the ultimate fate of the cosmos, remain the subject of active research and debate among physicists and cosmologists.

170. The concept of dark matter refers to a form of matter that does not emit, absorb or interact significantly with electromagnetic radiation, making it invisible and detectable only through its gravitational effects

Quantum Metaphysical Investigations : Exploring the Depths of Reality and Consciousness
Marino Baca-Carmona

on visible matter. Dark matter is believed to make up approximately 27% of the universe's total mass energy content, making it one of the most abundant components of the cosmos. Its existence was first inferred from observations of the rotation curves of galaxies and the gravitational lensing of light from distant objects. While the nature of dark matter remains unknown, several candidate particles have been proposed, including weakly interacting massive particles (WIMPs), axions, and sterile neutrinos. Understanding dark matter is essential to unraveling the mysteries of galactic dynamics, the formation of cosmic structures, and the nature of fundamental interactions.

171. The concept of the observer effect in quantum mechanics refers to the phenomenon in which the act of observing a quantum system affects its behavior. In classical physics, the observer is considered separate from the system being observed, but in quantum mechanics, the observer is inherently entangled with the system. When a measurement is made on a quantum system, such as the position or momentum of a particle, the act of measurement perturbs the system and alters its state. The observer effect raises profound questions about the nature of reality, the role of consciousness in quantum mechanics, and the limits of scientific observation.

172. The concept of nonlocality in quantum mechanics refers to the phenomenon in which particles can instantaneously influence the states of each other, regardless of the distance between them. This behavior violates classical notions of locality and suggests a deep interconnection between quantum entities. Nonlocality was demonstrated in the Bell tests, where measurements on entangled particles showed correlations

Quantum Metaphysical Investigations : Exploring the Depths of Reality and Consciousness
Marino Baca-Carmona

that cannot be explained by classical physics. The implications of nonlocality extend to quantum information theory, quantum teleportation, and fundamental debates about the nature of reality and causality.

173. The concept of decoherence refers to the process by which quantum systems lose their coherence and behave in a classical manner due to interactions with their surrounding environment. Decoherence leads to the suppression of quantum interference effects, such as superposition and entanglement, and results in the emergence of classical behavior at macroscopic scales. Decoherence plays a crucial role in the transition from quantum to classical realms and is essential for understanding the classical limit of quantum mechanics, as well as for the emergence of classical reality from quantum substrates.

174. The concept of quantum tunneling refers to the phenomenon in which particles can pass through potential energy barriers that classical mechanics would classically prohibit. In quantum mechanics, particles are described by wave functions that extend beyond classical limits, allowing them to penetrate through classically forbidden regions. Quantum tunneling has important applications in several fields, including semiconductor devices, nuclear fusion, and quantum computing. Theoretical and experimental studies of quantum tunneling provide insights into the fundamental nature of quantum mechanics and the behavior of matter at the quantum level.

175. The concept of spontaneous symmetry breaking refers to a phenomenon in physics in which a system exhibits a symmetric

Quantum Metaphysical Investigations : Exploring the Depths of Reality and Consciousness
Marino Baca-Carmona

configuration at high energies, but transitions to a low-energy state with broken symmetry. This transition results in the appearance of different terrestrial states and the spontaneous generation of massless Goldstone bosons. Spontaneous symmetry breaking plays a crucial role in many areas of physics, including particle physics, condensed matter physics, and cosmology. It is responsible for the generation of particle masses in the Standard Model of particle physics and the formation of cosmic structures in the early universe.

176. The concept of renormalization is a technique used in quantum field theory to address infinities that arise in calculations of physical quantities, such as particle interactions and field fluctuations. Renormalization involves systematically eliminating divergences by redefining parameters and absorbing them in finite, physically meaningful quantities. Renormalization has been successfully applied in several areas of theoretical physics, including quantum electrodynamics, the Standard Model of particle physics, and quantum chromodynamics. Although initially met with skepticism, renormalization has become a cornerstone of modern theoretical physics, providing a framework for making accurate predictions and understanding the fundamental forces of nature.

177. The concept of black hole evaporation, also known as Hawking radiation, refers to the theoretical prediction that black holes can emit thermal radiation due to quantum effects near the event horizon. This radiation is named after physicist Stephen Hawking, who first proposed its existence in 1974. According to Hawking's calculations, black holes should gradually lose mass over time as they emit radiation, eventually

leading to their complete evaporation. . The evaporation of black holes has profound implications for the information paradox, the fate of black holes, and the connections between gravity and quantum mechanics.

178. The concept of the holographic universe suggests that the information content of a three-dimensional region of space can be encoded on a two-dimensional surface, much like a hologram. This idea arises from theoretical considerations in quantum gravity and string theory, where black holes are believed to store information about their event horizons in a way consistent with holographic principles. The holographic universe hypothesis provides a novel perspective on questions related to the nature of space-time, the fundamental structure of reality, and the origin of entropy. Although speculative, the holographic universe hypothesis has stimulated new directions of research in theoretical physics and cosmology.

179. The concept of cosmic inflation proposes that the universe underwent a brief period of exponential expansion early in its existence, driven by a scalar field known as an inflaton. This rapid expansion smoothed the curvature of space, spread quantum fluctuations on cosmic scales, and seeded the primordial density fluctuations that eventually gave rise to the large-scale structure of the universe, such as galaxies and galaxy clusters. Cosmic inflation provides an elegant explanation for several key characteristics of the universe, including its homogeneity, isotropy, and flatness. However, the precise mechanisms that triggered inflation, as well as its implications for the multiverse and the ultimate fate of the cosmos, remain the subject of active research and debate among physicists and cosmologists.

Quantum Metaphysical Investigations : Exploring the Depths of Reality and Consciousness

Marino Baca-Carmona

180. The concept of dark energy refers to a mysterious form of energy that permeates all of space and drives the accelerated expansion of the universe. Dark energy is believed to make up about 68% of the universe's total energy density, making it the dominant component of the cosmos. Its existence was inferred from observations of distant supernovae and the large-scale distribution of galaxies. Despite its prevalence, the nature of dark energy remains one of the deepest mysteries in cosmology. Several theoretical models have been proposed to explain dark energy, including the cosmological constant, quintessence, and modifications of Einstein's theory of gravity. Understanding dark energy is essential to unraveling the ultimate fate of the universe and the underlying physics that governs cosmic evolution.

181. The concept of the cosmological constant, introduced by Albert Einstein in his field equations of general relativity, represents a constant energy density that is inherent to space itself. Initially proposed by Einstein to achieve a static universe model, the cosmological constant can also be interpreted as a form of dark energy that drives the accelerated expansion of the universe. The value of the cosmological constant remains one of the most intriguing puzzles in cosmology, with theoretical predictions differing from observed values by many orders of magnitude. Its presence has profound implications for the fate of the universe, the nature of dark energy, and the fundamental laws of physics.

182. The concept of loop quantum gravity is a theoretical framework that seeks to reconcile general relativity with quantum mechanics by quantizing the fabric of spacetime itself. In loop quantum

Quantum Metaphysical Investigations : Exploring the Depths of Reality and Consciousness
Marino Baca-Carmona

gravity, spacetime is discretized into tiny units or "loops," and the dynamics of these loops are governed by quantum principles. Loop quantum gravity provides a non-perturbative approach to quantum gravity, addressing the singularities present in classical general relativity and providing insights into the behavior of spacetime at the Planck scale. While still a work in progress, loop quantum gravity offers a promising avenue for understanding the quantum nature of spacetime and the fundamental structure of the universe.

    183. The concept of the string landscape arises from string theory, a theoretical framework that describes fundamental particles as vibrating strings rather than point particles. In string theory, the vacuum is not unique but instead consists of a vast landscape of possible vacua, each corresponding to different configurations of string vibrations and compactifications of extra dimensions. The string landscape gives rise to a multitude of possible universes with different physical constants, particle masses, and fundamental forces. This landscape has profound implications for the anthropic principle, the multiverse hypothesis, and the search for a unified theory of fundamental interactions.

    184. The concept of the information paradox arises from the conflict between the principles of quantum mechanics and general relativity in the context of black hole physics. According to quantum mechanics, information is conserved, meaning that the complete state of a system can be reconstructed from its quantum state at any given time. However, in the process of black hole formation and evaporation, it appears that information can be irretrievably lost, violating the principles of quantum mechanics. Resolving the information paradox is one of the most challenging problems in theoretical physics and may require a

Quantum Metaphysical Investigations : Exploring the Depths of Reality and Consciousness
Marino Baca-Carmona

deeper understanding of quantum gravity and the structure of black holes.

    185. The concept of the arrow of time refers to the asymmetry of time's directionality, where events unfold in a particular order from past to future. This asymmetry is manifested in various physical processes, such as the increase of entropy over time, the irreversibility of certain phenomena, and the apparent flow of time experienced by conscious observers. The arrow of time is intimately connected to the second law of thermodynamics, which states that the entropy of a closed system tends to increase over time, leading to the emergence of temporal asymmetry in macroscopic phenomena. Understanding the origin and nature of the arrow of time is a fundamental challenge in physics, philosophy, and cosmology.

    186. The concept of the Boltzmann brain paradox arises from certain cosmological models that predict the spontaneous formation of self-aware observers (Boltzmann brains) in a fluctuating universe. According to these models, fluctuations in the quantum vacuum may occasionally give rise to highly improbable configurations, such as fully formed brains with false memories, rather than the more typical evolution of complex structures through natural processes. The existence of Boltzmann brains raises profound questions about the nature of consciousness, the reliability of empirical observations, and the anthropic principle, which suggests that our observations of the universe are biased by our existence as observers.

    187. The concept of the Wheeler-DeWitt equation is a foundational equation in the field of quantum gravity that seeks to unify the principles of quantum mechanics and general relativity. Named after physicists John Archibald Wheeler and Bryce DeWitt, the Wheeler-

Quantum Metaphysical Investigations : <u>Exploring the Depths of Reality and Consciousness</u>
Marino Baca-Carmona

DeWitt equation is a constraint equation that describes the wave function of the universe, representing all possible configurations of spacetime geometry. Solving the Wheeler-DeWitt equation is a key goal of quantum cosmology, as it would provide insights into the quantum nature of the cosmos and the fundamental laws governing its evolution.

188. The concept of the holographic principle suggests that the information content of a three-dimensional region of space can be encoded on a two-dimensional surface, much like a hologram. This idea arises from theoretical considerations in quantum gravity and string theory, where black holes are believed to store information on their event horizons in a manner consistent with holographic principles. The holographic principle implies a deep connection between gravity, quantum mechanics, and information theory, with profound implications for our understanding of spacetime, black hole physics, and the fundamental structure of reality.

189. The concept of the multiverse refers to the hypothetical existence of multiple universes, each with its own set of physical laws, constants, and properties. The idea of a multiverse arises from various cosmological theories, including eternal inflation, string theory, and quantum mechanics. In these theories, different regions of space-time can undergo independent inflationary epochs, resulting in the formation of distinct "bubble" universes with different properties. The multiverse hypothesis has profound implications for our understanding of cosmology, the anthropic principle, and the nature of reality, but it remains speculative and difficult to test experimentally.

190. The concept of the many-worlds interpretation (MWI) of quantum mechanics proposes that every possible outcome of a quantum measurement occurs in a separate branch of the universe, leading to a

Quantum Metaphysical Investigations : Exploring the Depths of Reality and Consciousness
Marino Baca-Carmona

branching structure of reality. According to MWI, when a quantum system undergoes a measurement, the universe splits into multiple parallel realities, each corresponding to a different measurement outcome. MWI provides a deterministic and unitary interpretation of quantum mechanics, resolving the apparent randomness of quantum measurements. While MWI remains controversial and alternative interpretations exist, it offers a fascinating perspective on the nature of quantum reality and the structure of the universe.

191. The concept of quantum entanglement refers to the phenomenon in which two or more particles are correlated in such a way that the state of one particle instantaneously influences the state of the other particle(s), regardless of the distance between them. . This phenomenon arises from the principles of quantum mechanics, where particles can exist in superpositions of multiple states until a measurement is made, collapsing the wave function and determining their state. Quantum entanglement has been experimentally confirmed through numerous proofs of Bell's theorem and plays a crucial role in quantum information theory, quantum cryptography, and the basis of quantum mechanics.

192. The concept of the many-body problem refers to the challenge of accurately predicting the behavior of a system composed of multiple interacting particles. In classical mechanics, the equations of motion for individual particles can be solved directly using Newton's laws. However, when it comes to systems containing a large number of particles, such as gases, liquids or solids, the interactions between the particles become very complex, making it difficult to predict their collective behavior. The many-body problem is of central importance in

Quantum Metaphysical Investigations : Exploring the Depths of Reality and Consciousness
Marino Baca-Carmona

condensed matter physics, astrophysics, and plasma physics, where it is essential to understand the emergent properties of complex systems.

193. The concept of the Bose-Einstein condensate (BEC) is a state of matter that occurs at extremely low temperatures, near absolute zero, where a group of bosons (particles with integer spin) occupy the same quantum state. In a BEC, the wave functions of individual particles overlap, leading to macroscopic quantum phenomena such as superfluidity and coherence. BECs were first predicted by Satyendra Nath Bose and Albert Einstein in the 1920s and were experimentally realized in dilute atomic gases in the 1990s. Since then, BECs have become a powerful tool for studying quantum phenomena and They have applications in precision measurements, quantum information processing, and fundamental physics research.

194. The concept of the Pauli exclusion principle is a fundamental principle of quantum mechanics that states that no two identical fermions (particles with half-integer spin) can occupy the same quantum state simultaneously. This principle, formulated by Wolfgang Pauli in 1925, explains the stability of matter and the structure of atoms, where the electrons of an atom must occupy different quantum states with different quantum numbers. The Pauli exclusion principle has profound implications for the behavior of fermionic systems, including the formation of electron shells in atoms, the degeneracy pressure in white dwarf stars, and the properties of neutron stars.

195. The concept of the Casimir effect is a quantum mechanical phenomenon in which two closely spaced parallel conducting plates

Quantum Metaphysical Investigations : Exploring the Depths of Reality and Consciousness

Marino Baca-Carmona

experience an attractive force due to vacuum fluctuations of the electromagnetic field. This force arises from the confinement of virtual particles between the plates, which leads to a difference in vacuum energy density and results in a net force pushing the plates together. The Casimir effect was first predicted by Dutch physicist Hendrik Casimir in 1948 and has since been experimentally confirmed in various settings, including microelectromechanical systems (MEMS) and nanotechnology applications. Understanding the Casimir effect is essential for small-scale precision engineering and has implications for the design of nanoscale devices and quantum technologies.

196. The concept of superconductivity refers to the phenomenon in which certain materials exhibit zero electrical resistance and expel magnetic fields when cooled below a critical temperature. Superconductivity was first discovered in mercury by Dutch physicist Heike Kamerlingh Onnes in 1911, and subsequent research has revealed a rich variety of superconducting materials with diverse properties and applications. Superconductors are used in a wide range of technologies, including magnetic resonance imaging (MRI), particle accelerators, and power transmission lines. Understanding the mechanisms underlying superconductivity, such as Cooper pairing and the formation of superconducting condensates, is essential to advancing the field of condensed matter physics and developing practical applications of superconductors.

197. The concept of the uncertainty principle, formulated by Werner Heisenberg in 1927, states that certain pairs of physical properties, such as position and momentum, cannot be measured simultaneously with

Quantum Metaphysical Investigations : Exploring the Depths of Reality and Consciousness
Marino Baca-Carmona

arbitrary precision. Instead, there is a fundamental limit to the precision with which these properties can be known, given by the product of their uncertainties. The uncertainty principle arises from the wave-particle duality of quantum mechanics, where particles exhibit behavior similar to both the wave and the particle. It has profound implications for our understanding of the microscopic world and plays a central role in quantum mechanics, quantum field theory and modern technology such as electron microscopy and atomic spectroscopy.

198. The concept of topological insulators is a class of materials that behave as bulk insulators, but conduct electricity along their surface or edges due to topologically shielded states. Topological insulators are characterized by non-trivial band structures, where the electronic states exhibit a bulk energy gap and robust edge or surface states that are protected from disorder and imperfections. These materials have attracted significant interest due to their potential applications in spintronics, quantum computing, and low-power electronics. Understanding the topological properties of materials and engineering new topological phases is a rapidly growing area of research in condensed matter physics and materials science.

199. The concept of the photoelectric effect refers to the phenomenon in which electrons are ejected from the surface of a material when illuminated by electromagnetic radiation, such as visible light or ultraviolet radiation. The photoelectric effect was first discovered by Heinrich Hertz in 1887 and later explained by Albert Einstein in 1905 using the concept of quantized light energy (photons) and conservation of energy. The photoelectric effect has numerous applications, including

Quantum Metaphysical Investigations : Exploring the Depths of Reality and Consciousness
Marino Baca-Carmona

photovoltaic cells, photoemission spectroscopy, and photomultiplier tubes. Understanding the photoelectric effect is essential to advancing our knowledge of quantum mechanics and developing technologies to harness solar energy and detect light.

200. The concept of quantum field theory (QFT) is a theoretical framework that combines the principles of quantum mechanics with those of special relativity to describe the behavior of fundamental particles and fields. In QFT, particles are represented as excitations of quantum fields that permeate all of spacetime, and interactions between particles are described by the exchange of virtual particles. Quantum field theory provides a powerful framework for understanding the fundamental forces of nature, including electromagnetism, the weak nuclear force, the strong nuclear force, and gravity (in the context of quantum gravity). QFT has applications in particle physics, condensed matter physics, cosmology and quantum information theory, and is considered one of the most successful theories in modern physics.

201. The concept of gauge symmetry is a fundamental principle in theoretical physics, particularly in the formulation of gauge theories such as electromagnetism and the Standard Model of particle physics. Gauge symmetry refers to the invariance of physical laws under certain transformations called gauge transformations. In gauge theories, the Lagrangian that describes the dynamics of the fields is invariant under these transformations, leading to conservation laws and the emergence of gauge fields, such as the electromagnetic field or the gluon field. Gauge symmetry plays a crucial role in our understanding of the

fundamental forces of nature and the mathematical structures underlying modern theoretical physics.

202. The concept of the Higgs mechanism is a theoretical framework in particle physics that explains how elementary particles acquire mass. Proposed by physicists Peter Higgs, Robert Brout, and François Englert in the 1960s, the Higgs mechanism postulates the existence of a scalar field, called the Higgs field, that permeates all of spacetime. Interactions with this field give rise to mass for certain particles, such as the W and Z bosons in the Standard Model. The discovery of the Higgs boson by experiments at the Large Hadron Collider in 2012 confirmed the existence of the Higgs field and provided crucial evidence for the Higgs mechanism.

203. The concept of spontaneous symmetry breaking (SSB) is a phenomenon in physics in which the ground state of a system does not exhibit the symmetries present in the underlying laws of physics. Instead, the ground state breaks these symmetries, resulting in the appearance of distinct low-energy states with broken symmetry. Spontaneous symmetry breaking plays a crucial role in several areas of physics, including condensed matter physics, particle physics, and cosmology. It is responsible for the generation of particle masses in the Standard Model, the formation of cosmic structures in the early universe, and the appearance of phase transitions in many-body systems.

204. The concept of dark matter refers to a hypothetical form of matter that does not emit, absorb or reflect electromagnetic radiation, making it invisible to conventional telescopes. Dark matter is inferred from its

Quantum Metaphysical Investigations : <u>Exploring the Depths of Reality and Consciousness</u>
Marino Baca-Carmona

gravitational effects on visible matter and light, such as the rotation curves of galaxies, the dynamics of galaxy clusters, and the large-scale structure of the universe. Observational evidence suggests that dark matter makes up about 27% of the total mass-energy content of the cosmos, making it one of the most abundant components of the universe. Despite its prevalence, the nature of dark matter remains unknown and is a topic of active research in astrophysics and particle physics.

205. The concept of cosmic inflation is a theoretical framework in cosmology that proposes that the universe underwent rapid, exponential expansion in the early moments of its existence. This inflationary period, driven by a scalar field called an inflaton, is believed to have smoothed the curvature of space, stretched quantum fluctuations on cosmic scales, and seeded the primordial density fluctuations that ultimately gave rise to the large-scale structure of the universe, such as galaxies and galaxy clusters. Cosmic inflation provides an elegant explanation for several key features of the universe, including its homogeneity, isotropy, and flatness, and is supported by observational evidence such as cosmic microwave background radiation.

206. The concept of the anthropic principle is a philosophical and scientific principle that states that the observed properties of the universe must be compatible with the existence of observers, such as intelligent life. The anthropic principle arises from the apparent fine-tuning of the fundamental constants and parameters of the universe, which appear to be finely tuned to allow the emergence of life. There are several versions of the anthropic principle, ranging from the weak anthropic principle (which states that the universe must support the existence of observers)

to the strong anthropic principle (which suggests that the universe must have properties that support the existence of observers).

207. The concept of the many-worlds interpretation (MWI) of quantum mechanics is a controversial interpretation that suggests that each possible result of a quantum measurement occurs in a separate branch of the universe, leading to a branching structure of reality. According to MWI, when a quantum system is subjected to measurement, the universe is divided into multiple parallel realities, each of which corresponds to a different measurement result. MWI provides a deterministic and unitary interpretation of quantum mechanics, resolving the apparent randomness of quantum measurements. While MWI remains controversial and alternative interpretations exist, it offers a fascinating perspective on the nature of quantum reality and the structure of the universe.

208. The concept of supersymmetry (SUSY) is a theoretical framework in particle physics that postulates a symmetry between particles with integer spin (bosons) and particles with half-integer spin (fermions). Supersymmetry predicts the existence of new particles, known as superpartners, that have not yet been observed in experiments. These particles could provide a solution to several outstanding problems in particle physics, such as the hierarchy problem, the nature of dark matter, and the unification of fundamental forces. Despite extensive searches at particle colliders such as the Large Hadron Collider, direct evidence for supersymmetry remains elusive, leading some physicists to question its validity.

Quantum Metaphysical Investigations : Exploring the Depths of Reality and Consciousness

Marino Baca-Carmona

209. The concept of quantum gravity is a theoretical framework that seeks to reconcile the principles of quantum mechanics with those of general relativity, which describes gravity as the curvature of space-time. Quantum gravity is necessary to describe the behavior of matter and energy at the smallest scales, such as Planck length and Planck time, where quantum effects become significant. There are several approaches to quantum gravity, including string theory, loop quantum gravity, and causal dynamical triangulation, each with its own mathematical formalism and predictions. Understanding quantum gravity is essential to developing a unified theory of fundamental interactions and resolving theoretical inconsistencies between quantum mechanics and general relativity.

210. The concept of black hole thermodynamics is a theoretical framework that extends the laws of thermodynamics to black holes, which are regions of space-time with gravitational fields so strong that nothing, not even light, can escape from them. In the 1970s, physicist Stephen Hawking demonstrated that black holes can emit thermal radiation, now known as Hawking radiation, and have an associated temperature and entropy. This discovery established parallels between black holes and thermodynamic systems, leading to the formulation of laws of black hole mechanics analogous to the laws of thermodynamics. Black hole thermodynamics provides insights into the nature of black holes, their entropy, and their role in the universe.

211. The concept of the holographic principle suggests that the information content of a three-dimensional region of space can be encoded on a two-dimensional surface, much like a hologram. This idea

arises from theoretical considerations in quantum gravity and string theory, where black holes are believed to store information about their event horizons in a way consistent with holographic principles. The holographic principle implies a deep connection between gravity, quantum mechanics and information theory, with profound implications for our understanding of space-time, the physics of black holes and the fundamental structure of reality.

212. The concept of the string landscape arises from string theory, a theoretical framework that describes fundamental particles as vibrating strings rather than point particles. In string theory, the vacuum is not single, but consists of a vast landscape of possible vacuums, each corresponding to different configurations of string vibrations and compactifications of extra dimensions. The string landscape gives rise to a multitude of possible universes with different physical constants, particle masses and fundamental forces. This landscape has profound implications for the anthropic principle, the multiverse hypothesis, and the search for a unified theory of fundamental interactions. The existence of the string landscape suggests that our universe may be just one of many possible configurations within a vast multiverse, each with its own unique properties and laws of physics. Exploring the string landscape and understanding its implications for cosmology, particle physics, and the nature of reality is a central goal of research in theoretical physics.

213. The concept of quantum entanglement refers to the phenomenon where two or more particles become correlated in such a way that the state of one particle instantaneously influences the state of

Quantum Metaphysical Investigations : Exploring the Depths of Reality and Consciousness

Marino Baca-Carmona

the other(s), regardless of the distance between them. This phenomenon arises from the principles of quantum mechanics, where particles can exist in superpositions of multiple states until a measurement is made, collapsing the wave function and determining their state. Quantum entanglement has been experimentally confirmed through numerous tests of Bell's theorem and plays a crucial role in quantum information theory, quantum cryptography, and the foundation of quantum mechanics.

214. The concept of the Wheeler-DeWitt equation is a foundational equation in the field of quantum gravity that seeks to unify the principles of quantum mechanics and general relativity. Named after physicists John Archibald Wheeler and Bryce DeWitt, the Wheeler-DeWitt equation is a constraint equation that describes the wave function of the universe, representing all possible configurations of spacetime geometry. Solving the Wheeler-DeWitt equation is a key goal of quantum cosmology, as it would provide insights into the quantum nature of the cosmos and the fundamental laws governing its evolution.

215. The concept of loop quantum gravity is a theoretical framework that seeks to reconcile general relativity with quantum mechanics by quantizing the fabric of spacetime itself. In loop quantum gravity, spacetime is discretized into tiny units or "loops," and the dynamics of these loops are governed by quantum principles. Loop quantum gravity provides a non-perturbative approach to quantum gravity, addressing the singularities present in classical general relativity and providing insights into the behavior of spacetime at the Planck scale. While still a work in progress, loop quantum gravity offers a promising avenue for understanding the quantum nature of spacetime and the fundamental structure of the universe.

Quantum Metaphysical Investigations : Exploring the Depths of Reality and Consciousness
Marino Baca-Carmona

216. The concept of the Boltzmann brain paradox arises from certain cosmological models that predict the spontaneous formation of self-aware observers (Boltzmann brains) in a fluctuating universe. According to these models, fluctuations in the quantum vacuum may occasionally give rise to highly improbable configurations, such as fully formed brains with false memories, rather than the more typical evolution of complex structures through natural processes. The existence of Boltzmann brains raises profound questions about the nature of consciousness, the reliability of empirical observations, and the anthropic principle, which suggests that our observations of the universe are biased by our existence as observers.

217. The concept of quantum field theory (QFT) is a theoretical framework that combines the principles of quantum mechanics with those of special relativity to describe the behavior of fundamental particles and fields. In QFT, particles are represented as excitations of quantum fields that permeate all of spacetime, and interactions between particles are described by exchange of virtual particles. Quantum field theory provides a powerful framework for understanding the fundamental forces of nature, including electromagnetism, weak nuclear force, strong nuclear force, and gravity (in the context of quantum gravity). QFT has applications in particle physics, condensed matter physics, cosmology, and quantum information theory, and is considered one of the most successful theories in modern physics.

218. The concept of topological insulators is a class of materials that behave as insulators in their bulk but conduct electricity along their surface or edges due to topologically protected states. Topological insulators are characterized by non-trivial band structures, where the electronic states exhibit a bulk energy gap and robust edge or surface

Quantum Metaphysical Investigations : <u>Exploring the Depths of Reality and Consciousness</u>

Marino Baca-Carmona

states that are protected against disorder and imperfections. These materials have attracted significant interest due to their potential applications in spintronics, quantum computing, and low-power electronics. Understanding the topological properties of materials and engineering new topological phases is a rapidly growing area of research in condensed matter physics and materials science.

219. The concept of the Casimir effect is a quantum mechanical phenomenon where two closely spaced parallel conducting plates experience an attractive force due to the vacuum fluctuations of the electromagnetic field. This force arises from the confinement of virtual particles between the plates, leading to a difference in vacuum energy density and resulting in a net force that pushes the plates together. The Casimir effect was first predicted by Dutch physicist Hendrik Casimir in 1948 and has since been experimentally confirmed in various settings, including micro-electromechanical systems (MEMS) and nanotechnology applications. Understanding the Casimir effect is essential for precision engineering at small scales and has implications for the design of nanoscale devices and quantum technologies.

220. The concept of superconductivity refers to the phenomenon where certain materials exhibit zero electrical resistance and expel magnetic fields when cooled below a critical temperature. Superconductivity was first discovered in mercury by Dutch physicist Heike Kamerlingh Onnes in 1911, and subsequent research has revealed a rich variety of superconducting materials with diverse properties and applications. Superconductors are used in a wide range of technologies, including magnetic resonance imaging (MRI), particle accelerators, and power transmission lines. Understanding the mechanisms underlying superconductivity, such as Cooper pairing and the formation of

Quantum Metaphysical Investigations : <u>Exploring the Depths of Reality and Consciousness</u>
Marino Baca-Carmona

superconducting condensates, is essential for advancing the field of condensed matter physics and developing practical applications of superconductors.

221. The concept of cosmic microwave background radiation (CMB) refers to the thermal radiation left over from the Big Bang that permeates the entire universe. CMB radiation was first predicted by George Gamow, Ralph Alpher and Robert Herman in the 1940s and was accidentally discovered by Arno Penzias and Robert Wilson in 1965. It is crucial evidence supporting the Big Bang model of cosmology and providing information about the early history and evolution of the universe. CMB radiation is nearly isotropic and exhibits small fluctuations in temperature, which have been mapped in detail by experiments such as the Cosmic Background Explorer (COBE), the Wilkinson Microwave Anisotropy Probe (WMAP) and the Planck satellite.

222. The concept of black hole evaporation, also known as Hawking radiation, refers to physicist Stephen Hawking's theoretical prediction that black holes can emit thermal radiation due to quantum effects near their event horizons. According to Hawking's theory, pairs of virtual particles and antiparticles can be created near the event horizon of a black hole, with one particle falling into the black hole and the other escaping into space. Over time, this process leads to a net loss of mass and energy from the black hole, causing it to evaporate and eventually disappear. While black hole evaporation has not yet been directly observed, it is a fundamental prediction of quantum field theory in curved spacetime and has profound implications for black hole

thermodynamics and the fate of information in the universe. quantum gravity.

223. The concept of the multiverse refers to the idea that our universe may be just one of many universes that exist parallel to each other, each with its own unique properties and laws of physics. The multiverse hypothesis arises from several cosmological models, such as eternal inflation, string theory, and quantum mechanics, which suggest that our universe may be just one of countless "bubbles" or regions within a larger multiversal structure. The existence of a multiverse could potentially explain the apparent adjustment of nature's fundamental constants, the origin of the cosmological constant problem, and the nature of dark energy. While the multiverse hypothesis remains speculative and controversial, it has sparked significant interest among physicists and cosmologists as a possible explanation for the observed features of our universe.

224. The concept of string theory is a theoretical framework in theoretical physics that attempts to describe the fundamental components of the universe as one-dimensional "strings" rather than point particles. In string theory, the different modes of vibration of these strings correspond to different types of particles and interactions, providing a unified description of gravity, electromagnetism, and the other fundamental forces of nature. String theory also predicts the existence of additional dimensions beyond the familiar three spatial dimensions and one time dimension, which can be compacted to small scales. While string theory has yet to be experimentally confirmed, it

remains a leading candidate for a theory of quantum gravity and a potential framework for unifying all known physical interactions.

225. The concept of the holographic principle suggests that the information content of a three-dimensional region of space can be encoded on a two-dimensional surface, much like a hologram. This idea arises from theoretical considerations in quantum gravity and string theory, where black holes are believed to store information about their event horizons in a way consistent with holographic principles. The holographic principle implies a deep connection between gravity, quantum mechanics and information theory, with profound implications for our understanding of space-time, the physics of black holes and the fundamental structure of reality.

226. The concept of quantum entanglement refers to the phenomenon in which two or more particles are correlated in such a way that the state of one particle instantaneously influences the state of the other particle(s), regardless of the distance between them. . This phenomenon arises from the principles of quantum mechanics, where particles can exist in superpositions of multiple states until a measurement is made, collapsing the wave function and determining their state. Quantum entanglement has been experimentally confirmed through numerous proofs of Bell's theorem and plays a crucial role in quantum information theory, quantum cryptography, and the basis of quantum mechanics.

227. The concept of the Wheeler-DeWitt equation is a fundamental equation in the field of quantum gravity that seeks to unify the principles of quantum mechanics and general relativity. Named after physicists

Quantum Metaphysical Investigations : Exploring the Depths of Reality and Consciousness
Marino Baca-Carmona

John Archibald Wheeler and Bryce DeWitt, the Wheeler-DeWitt equation is a constraint equation that describes the wave function of the universe, representing all possible configurations of space-time geometry. Solving the Wheeler-DeWitt equation is a key goal of quantum cosmology, as it would provide insights into the quantum nature of the cosmos and the fundamental laws governing its evolution.

228. The concept of loop quantum gravity is a theoretical framework that seeks to reconcile general relativity with quantum mechanics by quantifying the fabric of spacetime itself. In loop quantum gravity, spacetime is discretized into small units or "loops," and the dynamics of these loops are governed by quantum principles. Loop quantum gravity provides a non-perturbative approach to quantum gravity, addressing the singularities present in classical general relativity and providing insights into the behavior of spacetime on the Planck scale. While still a work in progress, loop quantum gravity offers a promising avenue for understanding the quantum nature of spacetime and the fundamental structure of the universe.

229. The concept of the Boltzmann brain paradox arises from certain cosmological models that predict the spontaneous formation of self-conscious observers (Boltzmann brains) in a fluctuating universe. According to these models, fluctuations in the quantum vacuum can occasionally give rise to highly improbable configurations, such as fully formed brains with false memories, rather than the more typical evolution of complex structures through natural processes. The existence of Boltzmann brains raises profound questions about the nature of consciousness, the reliability of empirical observations, and the

Quantum Metaphysical Investigations : <u>Exploring the Depths of Reality and Consciousness</u>
Marino Baca-Carmona

anthropic principle, suggesting that our observations of the universe are biased by our existence as observers.

230. The concept of quantum field theory (QFT) is a theoretical framework that combines the principles of quantum mechanics with those of special relativity to describe the behavior of fundamental particles and fields. In QFT, particles are represented as excitations of quantum fields that permeate all of spacetime, and interactions between particles are described by the exchange of virtual particles. Quantum field theory provides a powerful framework for understanding the fundamental forces of nature, including electromagnetism, the weak nuclear force, the strong nuclear force, and gravity (in the context of quantum gravity). QFT has applications in particle physics, condensed matter physics, cosmology and quantum information theory, and is considered one of the most successful theories in modern physics.

231. The concept of topological insulators is a class of materials that behave as bulk insulators, but conduct electricity along their surface or edges due to topologically shielded states. Topological insulators are characterized by non-trivial band structures, where the electronic states exhibit a bulk energy gap and robust edge or surface states that are protected from disorder and imperfections. These materials have attracted significant interest due to their potential applications in spintronics, quantum computing, and low-power electronics. Understanding the topological properties of materials and engineering new topological phases is a rapidly growing area of research in condensed matter physics and materials science.

Quantum Metaphysical Investigations : Exploring the Depths of Reality and Consciousness

Marino Baca-Carmona

232. The concept of The Casimir effect is a quantum mechanical phenomenon in which two closely spaced parallel conducting plates experience an attractive force due to vacuum fluctuations of the electromagnetic field. This force arises from the confinement of virtual particles between the plates, which leads to a difference in vacuum energy density and results in a net force pushing the plates together. The Casimir effect was first predicted by Dutch physicist Hendrik Casimir in 1948 and has since been experimentally confirmed in various settings, including microelectromechanical systems (MEMS) and nanotechnology applications. Understanding the Casimir effect is essential for small-scale precision engineering and has implications for the design of nanoscale devices and quantum technologies.

233. The concept of superconductivity refers to the phenomenon in which certain materials exhibit zero electrical resistance and expel magnetic fields when cooled below a critical temperature. Superconductivity was first discovered in mercury by Dutch physicist Heike Kamerlingh Onnes in 1911, and subsequent research has revealed a rich variety of superconducting materials with diverse properties and applications. Superconductors are used in a wide range of technologies, including magnetic resonance imaging (MRI), particle accelerators, and power transmission lines. Understanding the mechanisms underlying superconductivity, such as Cooper pairing and the formation of superconducting condensates, is essential to advancing the field of condensed matter physics and developing practical applications of superconductors.

Quantum Metaphysical Investigations : Exploring the Depths of Reality and Consciousness
Marino Baca-Carmona

234. The concept of quantum tunneling is a phenomenon in quantum mechanics in which a particle penetrates through a classically forbidden energy barrier despite lacking sufficient energy to overcome it. This phenomenon arises from the wave nature of particles and is governed by the principles of the Heisenberg uncertainty principle. Quantum tunneling plays a crucial role in several physical processes, including nuclear fusion in stars, scanning tunneling microscopy in nanotechnology, and the operation of tunneling diodes in electronics. Understanding quantum tunneling is essential to explaining phenomena at the atomic and subatomic scale and has practical applications in technology and engineering.

235. The concept of quantum decoherence is a phenomenon in quantum mechanics in which the coherence and superposition of quantum states are lost due to interactions with the environment. Quantum systems become entangled with their environment, leading to the suppression of interference effects and the emergence of classical behavior. Quantum decoherence plays an important role in the transition from the quantum realm to the classical realm and is responsible for the apparent collapse of the wave function in measurements. Understanding quantum decoherence is essential for the development of quantum computing, quantum information processing, and other quantum technologies, where maintaining coherence is crucial to performing complex quantum operations.

236. The concept of quantum computing is a revolutionary approach to computing that uses the principles of quantum mechanics to perform calculations exponentially faster than classical computers for certain

tasks. In a quantum computer, information is encoded into quantum bits or qubits, which can exist in superpositions of multiple states and be entangled with each other. Quantum algorithms take advantage of these quantum properties to solve problems such as factoring large numbers, simulating quantum systems, and optimizing complex functions more efficiently than classical algorithms. Quantum computing has the potential to revolutionize fields such as cryptography, materials science, and artificial intelligence, but significant technical challenges remain in building practical quantum computers with sufficiently low error rates and scalability.

237. The concept of quantum teleportation is a quantum communication protocol that allows the transfer of quantum information from one place to another, without physically transporting the underlying particles. Quantum teleportation relies on the principles of entanglement and quantum measurement to transmit the state of a qubit from a sender to a receiver over long distances. While quantum teleportation cannot violate the speed of light or enable faster-than-light communication, it has applications in quantum cryptography, quantum networks, and distributed quantum computing. Understanding quantum teleportation is essential to developing secure communication protocols and building a quantum Internet for the future.

238. The concept of quantum entanglement exchange is an extension of quantum teleportation that allows the creation of entanglements between particles that have never directly interacted. In quantum entanglement exchange, two pairs of entangled particles, AB and CD, are generated separately and then subjected to joint measurements. By performing

measurements on particles B and C and exploiting the correlations between them, entanglement between particles A and D can be established, even though they have never directly interacted. Quantum entanglement sharing has applications in quantum communication, quantum key distribution, and distributed quantum computing, and is an essential technique for building large-scale quantum networks.

239. The concept of quantum supremacy refers to the milestone reached by a quantum computer when it can perform a specific task faster than the best classical computers available. This task, often chosen to be computationally intractable for classical computers, serves as a demonstration of the potential computational superiority of quantum systems. Achieving quantum supremacy requires not only building a quantum computer with sufficiently low error rates and qubit coherence times, but also designing and implementing quantum algorithms that take advantage of the unique properties of quantum mechanics. While quantum supremacy has yet to be definitively demonstrated, recent advances in quantum hardware and algorithm development have brought this milestone within reach of current research efforts.

240. The concept of quantum cryptography is a field of quantum information science that uses the principles of quantum mechanics to protect communication channels against eavesdropping and interception. Quantum cryptography protocols, such as quantum key distribution (QKD), rely on properties of quantum states, such as superposition and entanglement, to securely generate and distribute cryptographic keys between remote parties. Quantum cryptography offers provably secure methods for key distribution, ensuring the confidentiality and integrity

of transmitted data even in the presence of powerful adversaries with access to quantum computers. Quantum cryptography has applications in secure communication networks, financial transactions, and government and military communications, and is a key technology for ensuring privacy in the quantum age.

241. The concept of the many-worlds interpretation (MWI) is an interpretation of quantum mechanics that suggests that all possible results of a quantum measurement correspond to a separate and divergent branch of the universe. According to MWI, every time a quantum system undergoes a measurement or interaction, the universe splits into multiple parallel realities, each of which corresponds to a different result of the measurement. In this interpretation, the wave function never collapses, but rather all possible outcomes coexist in separate branches of the multiverse. MWI has profound implications for our understanding of probability, determinism, and the nature of reality, but remains a controversial and speculative interpretation of quantum mechanics.

242. The concept of quantum nonlocality refers to the phenomenon in which entangled quantum particles can exhibit correlations that cannot be explained by classical physics, even when separated by large distances. According to quantum mechanics, measurements made on one entangled particle instantly affect the state of the other particle, regardless of the distance between them, violating the principle of local realism. Quantum nonlocality has been confirmed experimentally through proofs of Bell's theorem and plays a central role in the

Quantum Metaphysical Investigations : <u>Exploring the Depths of Reality and Consciousness</u>
Marino Baca-Carmona

foundations of quantum mechanics, as well as in quantum information theory and quantum cryptography.

243. The concept of quantum chaos refers to the study of chaotic behavior in quantum systems, where small changes in initial conditions can lead to dramatically different results over time. While classical chaos arises from nonlinear dynamics in classical systems, quantum chaos involves the interaction of quantum mechanics and chaos theory in systems with discrete energy levels. Quantum chaos has applications in several fields, including nuclear physics, condensed matter physics, and quantum computing, and provides insights into the behavior of complex quantum systems and the limits of classical predictability.

244. The concept of the black hole information paradox arises from the conflict between the principles of quantum mechanics and classical black hole theory. According to quantum mechanics, information is always conserved, meaning that the complete physical state of a system can be determined from its initial conditions and its evolution over time. However, classical black holes are predicted to destroy any information that falls into them, leading to a violation of quantum principles. Resolving the black hole information paradox is a fundamental challenge in theoretical physics and may require a deeper understanding of the quantum nature of spacetime and the behavior of black hole horizons.

245. The concept of the holographic principle suggests that the information content of a three-dimensional region of space can be encoded on a two-dimensional surface, much like a hologram. This idea

Quantum Metaphysical Investigations : Exploring the Depths of Reality and Consciousness
Marino Baca-Carmona

arises from theoretical considerations in quantum gravity and string theory, where black holes are believed to store information about their event horizons in a way consistent with holographic principles. The holographic principle implies a deep connection between gravity, quantum mechanics and information theory, with profound implications for our understanding of space-time, the physics of black holes and the fundamental structure of reality.

246. The concept of quantum supremacy refers to the milestone reached by a quantum computer when it can perform a specific task faster than the best classical computers available. This task, often chosen to be computationally intractable for classical computers, serves as a demonstration of the potential computational superiority of quantum systems. Achieving quantum supremacy requires not only building a quantum computer with sufficiently low error rates and qubit coherence times, but also designing and implementing quantum algorithms that take advantage of the unique properties of quantum mechanics. While quantum supremacy has yet to be definitively demonstrated, recent advances in quantum hardware and algorithm development have brought this milestone within reach of current research efforts.

247. The concept of quantum key distribution (QKD) is a quantum cryptography protocol that allows two parties to securely exchange cryptographic keys over a potentially insecure communication channel. QKD relies on quantum mechanical principles, such as the no-cloning theorem and the uncertainty principle, to ensure that any attempt to eavesdrop on key exchanges is detected by legitimate parties. Quantum key distribution offers likely secure methods for key distribution,

Quantum Metaphysical Investigations : Exploring the Depths of Reality and Consciousness
Marino Baca-Carmona

ensuring the confidentiality and integrity of transmitted data even in the presence of powerful adversaries with access to quantum computers. QKD has applications in secure communication networks, financial transactions, and government and military communications, and is a key technology for ensuring privacy in the quantum era.

248. The concept of quantum teleportation is a quantum communication protocol that allows the transfer of quantum information from one place to another, without physically transporting the underlying particles. Quantum teleportation relies on the principles of entanglement and quantum measurement to transmit the state of a qubit from a sender to a receiver over long distances. While quantum teleportation cannot violate the speed of light or enable faster-than-light communication, it has applications in quantum cryptography, quantum networks, and distributed quantum computing. Understanding quantum teleportation is essential to developing secure communication protocols and building a quantum Internet for the future.

249. The concept of quantum entanglement exchange is an extension of quantum teleportation that allows the creation of entanglements between particles that have never directly interacted. In quantum entanglement exchange, two pairs of entangled particles, AB and CD, are generated separately and then subjected to joint measurements. By performing measurements on particles B and C and exploiting the correlations between them, entanglement between particles A and D can be established, even though they have never directly interacted. Quantum entanglement sharing has applications in quantum communication,

Quantum Metaphysical Investigations : Exploring the Depths of Reality and Consciousness

Marino Baca-Carmona

quantum key distribution, and distributed quantum computing, and is an essential technique for building large-scale quantum networks.

250. The concept of quantum error correction is a set of techniques and protocols designed to protect quantum information from errors and decoherence induced by noise and imperfections in quantum hardware. Quantum error correction is based on the encoding of quantum states into larger quantum codes that are robust against errors and can be efficiently corrected through error detection and correction procedures. Quantum error correction is crucial for building fault-tolerant quantum computers capable of reliably performing complex quantum algorithms, as well as for implementing secure quantum communication protocols and protecting quantum information from eavesdropping and tampering. Understanding quantum error correction is essential to overcoming the technical challenges of building practical quantum technologies and realizing the full potential of quantum computing and quantum communication.

251. The concept of quantum annealing is a computational optimization technique that takes advantage of quantum mechanics to find solutions to complex optimization problems more efficiently than classical algorithms. In quantum annealing, a quantum system is gradually cooled from a high-energy state to a low-energy state, allowing it to explore the solution space of an optimization problem and settle into a low-energy configuration corresponding to the solution. optimal. Quantum annealing has applications in several fields, including combinatorial optimization, machine learning, and materials science, and is being pursued as a

promising approach to solve large-scale optimization problems that are intractable for classical computers.

252. The concept of quantum machine learning is an emerging field at the intersection of quantum computing and machine learning that seeks to develop algorithms and techniques to harness quantum resources to enhance the capabilities of machine learning systems. Quantum machine learning algorithms aim to exploit the unique properties of quantum mechanics, such as superposition and entanglement, to perform tasks such as data classification, clustering, regression, and reinforcement learning more efficiently than classical algorithms. Quantum machine learning has the potential to revolutionize fields such as drug discovery, financial modeling, and natural language processing by enabling the processing of large-scale data sets and the discovery of complex patterns and correlations that are beyond the reach of classic machine learning techniques.

253. The concept of quantum-inspired optimization algorithms refers to classical optimization algorithms that are inspired by the principles and techniques of quantum computing. These algorithms aim to mimic the behavior of quantum annealing or adiabatic quantum computing, taking advantage of concepts such as quantum tunneling, quantum parallelism, and quantum superposition to explore solution spaces more efficiently than traditional optimization algorithms. Quantum-inspired optimization algorithms include methods such as quantum annealing-inspired algorithms, quantum-inspired genetic algorithms, and quantum-inspired particle swarm optimization. While these algorithms do not require real quantum hardware to run, they seek to exploit quantum behavior in

Quantum Metaphysical Investigations : Exploring the Depths of Reality and Consciousness
Marino Baca-Carmona

classical systems to solve complex optimization problems more effectively. Quantum-inspired optimization algorithms have applications in various fields, such as logistics, finance, and machine learning, where finding optimal solutions to large-scale optimization problems is essential to improve efficiency and performance.

254. The concept of quantum supremacy refers to the milestone reached by a quantum computer when it can perform a specific task faster than the best classical computers available. This task, often chosen to be computationally intractable for classical computers, serves as a demonstration of the potential computational superiority of quantum systems. Achieving quantum supremacy requires not only building a quantum computer with sufficiently low error rates and qubit coherence times, but also designing and implementing quantum algorithms that take advantage of the unique properties of quantum mechanics. While quantum supremacy has yet to be definitively demonstrated, recent advances in quantum hardware and algorithm development have brought this milestone within reach of current research efforts.

255. The concept of quantum key distribution (QKD) is a quantum cryptography protocol that allows two parties to securely exchange cryptographic keys over a potentially insecure communication channel. QKD relies on quantum mechanical principles, such as the no-cloning theorem and the uncertainty principle, to ensure that any attempt to eavesdrop on key exchanges is detected by legitimate parties. Quantum key distribution offers likely secure methods for key distribution, ensuring the confidentiality and integrity of transmitted data even in the presence of powerful adversaries with access to quantum computers.

Quantum Metaphysical Investigations : <u>Exploring the Depths of Reality and Consciousness</u>
Marino Baca-Carmona

QKD has applications in secure communication networks, financial transactions, and government and military communications, and is a key technology for ensuring privacy in the quantum era.

256. The concept of quantum teleportation is a quantum communication protocol that allows the transfer of quantum information from one place to another, without physically transporting the underlying particles. Quantum teleportation relies on the principles of entanglement and quantum measurement to transmit the state of a qubit from a sender to a receiver over long distances. While quantum teleportation cannot violate the speed of light or enable faster-than-light communication, it has applications in quantum cryptography, quantum networks, and distributed quantum computing. Understanding quantum teleportation is essential to developing secure communication protocols and building a quantum Internet for the future.

257. The concept of quantum entanglement exchange is an extension of quantum teleportation that allows the creation of entanglements between particles that have never directly interacted. In quantum entanglement exchange, two pairs of entangled particles, AB and CD, are generated separately and then subjected to joint measurements. By performing measurements on particles B and C and exploiting the correlations between them, entanglement between particles A and D can be established, even though they have never directly interacted. Quantum entanglement sharing has applications in quantum communication, quantum key distribution, and distributed quantum computing, and is an essential technique for building large-scale quantum networks.

Quantum Metaphysical Investigations : Exploring the Depths of Reality and Consciousness

Marino Baca-Carmona

258. The concept of quantum error correction is a set of techniques and protocols designed to protect quantum information from errors and decoherence induced by noise and imperfections in quantum hardware. Quantum error correction is based on the encoding of quantum states into larger quantum codes that are robust against errors and can be efficiently corrected through error detection and correction procedures. Quantum error correction is crucial for building fault-tolerant quantum computers capable of reliably performing complex quantum algorithms, as well as for implementing secure quantum communication protocols and protecting quantum information from eavesdropping and tampering. Understanding quantum error correction is essential to overcoming the technical challenges of building practical quantum technologies and realizing the full potential of quantum computing and quantum communication.

259. The concept of quantum annealing is a computational optimization technique that takes advantage of quantum mechanics to find solutions to complex optimization problems more efficiently than classical algorithms. In quantum annealing, a quantum system is gradually cooled from a high-energy state to a low-energy state, allowing it to explore the solution space of an optimization problem and settle into a low-energy configuration corresponding to the solution. optimal. Quantum annealing has applications in several fields, including combinatorial optimization, machine learning, and materials science, and is being pursued as a promising approach to solve large-scale optimization problems that are intractable for classical computers.

Quantum Metaphysical Investigations : Exploring the Depths of Reality and Consciousness
Marino Baca-Carmona

260. The concept of quantum machine learning is an emerging field at the intersection of quantum computing and machine learning that seeks to develop algorithms and techniques to harness quantum resources to enhance the capabilities of machine learning systems. Quantum machine learning algorithms aim to exploit the unique properties of quantum mechanics, such as superposition and entanglement, to perform tasks such as data classification, clustering, regression, and reinforcement learning more efficiently than classical algorithms. Quantum machine learning has the potential to revolutionize fields such as drug discovery, financial modeling, and natural language processing by enabling the processing of large-scale data sets and the discovery of complex patterns and correlations that are beyond the reach of classic machine learning techniques.

261. The concept of quantum supremacy refers to the milestone reached by a quantum computer when it can perform a specific task faster than the best classical computers available. This task, often chosen to be computationally intractable for classical computers, serves as a demonstration of the potential computational superiority of quantum systems. Achieving quantum supremacy requires not only building a quantum computer with sufficiently low error rates and qubit coherence times, but also designing and implementing quantum algorithms that take advantage of the unique properties of quantum mechanics. While quantum supremacy has yet to be definitively demonstrated, recent advances in quantum hardware and algorithm development have brought this milestone within reach of current research efforts.

Quantum Metaphysical Investigations : Exploring the Depths of Reality and Consciousness

Marino Baca-Carmona

262. The concept of quantum key distribution (QKD) is a quantum cryptography protocol that allows two parties to securely exchange cryptographic keys over a potentially insecure communication channel. QKD relies on quantum mechanical principles, such as the no-cloning theorem and the uncertainty principle, to ensure that any attempt to eavesdrop on key exchanges is detected by legitimate parties. Quantum key distribution offers likely secure methods for key distribution, ensuring the confidentiality and integrity of transmitted data even in the presence of powerful adversaries with access to quantum computers. QKD has applications in secure communication networks, financial transactions, and government and military communications, and is a key technology for ensuring privacy in the quantum era.

263. The concept of quantum teleportation is a quantum communication protocol that allows the transfer of quantum information from one place to another, without physically transporting the underlying particles. Quantum teleportation relies on the principles of entanglement and quantum measurement to transmit the state of a qubit from a sender to a receiver over long distances. While quantum teleportation cannot violate the speed of light or enable faster-than-light communication, it has applications in quantum cryptography, quantum networks, and distributed quantum computing. Understanding quantum teleportation is essential to developing secure communication protocols and building a quantum Internet for the future.

264. The concept of quantum entanglement exchange is an extension of quantum teleportation that allows the creation of entanglements between particles that have never directly interacted. In quantum entanglement

exchange, two pairs of entangled particles, AB and CD, are generated separately and then subjected to joint measurements. By performing measurements on particles B and C and exploiting the correlations between them, entanglement between particles A and D can be established, even though they have never directly interacted. Quantum entanglement sharing has applications in quantum communication, quantum key distribution, and distributed quantum computing, and is an essential technique for building large-scale quantum networks.

265. The concept of quantum error correction is a set of techniques and protocols designed to protect quantum information from errors and decoherence induced by noise and imperfections in quantum hardware. Quantum error correction is based on the encoding of quantum states into larger quantum codes that are robust against errors and can be efficiently corrected through error detection and correction procedures. Quantum error correction is crucial for building fault-tolerant quantum computers capable of reliably performing complex quantum algorithms, as well as for implementing secure quantum communication protocols and protecting quantum information from eavesdropping and tampering. Understanding quantum error correction is essential to overcoming the technical challenges of building practical quantum technologies and realizing the full potential of quantum computing and quantum communication.

266. The concept of quantum annealing is a computational optimization technique that takes advantage of quantum mechanics to find solutions to complex optimization problems more efficiently than classical algorithms. In quantum annealing, a quantum system is gradually cooled

Quantum Metaphysical Investigations : Exploring the Depths of Reality and Consciousness
Marino Baca-Carmona

from a high-energy state to a low-energy state, allowing it to explore the solution space of an optimization problem and settle into a low-energy configuration corresponding to the solution. optimal. Quantum annealing has applications in several fields, including combinatorial optimization, machine learning, and materials science, and is being pursued as a promising approach to solve large-scale optimization problems that are intractable for classical computers.

267. The concept of quantum machine learning is an emerging field at the intersection of quantum computing and machine learning that seeks to develop algorithms and techniques to harness quantum resources to enhance the capabilities of machine learning systems. Quantum machine learning algorithms aim to exploit the unique properties of quantum mechanics, such as superposition and entanglement, to perform tasks such as data classification, clustering, regression, and reinforcement learning more efficiently than classical algorithms. Quantum machine learning has the potential to revolutionize fields such as drug discovery, financial modeling, and natural language processing by enabling the processing of large-scale data sets and the discovery of complex patterns and correlations that are beyond the reach of classic machine learning techniques.

268. The concept of quantum-inspired optimization algorithms refers to classical optimization algorithms that are inspired by the principles and techniques of quantum computing. These algorithms aim to mimic the behavior of quantum annealing or adiabatic quantum computing, taking advantage of concepts such as quantum tunneling, quantum parallelism, and quantum superposition to explore solution spaces more efficiently

than traditional optimization algorithms. Quantum-inspired optimization algorithms include methods such as quantum annealing-inspired algorithms, quantum-inspired genetic algorithms, and quantum-inspired particle swarm optimization. While these algorithms do not require real quantum hardware to run, they seek to exploit quantum behavior in classical systems to solve complex optimization problems more effectively. Quantum-inspired optimization algorithms have applications in various fields, such as logistics, finance, and machine learning, where finding optimal solutions to large-scale optimization problems is essential to improve efficiency and performance.

269. The concept of quantum supremacy refers to the milestone reached by a quantum computer when it can perform a specific task faster than the best classical computers available. This task, often chosen to be computationally intractable for classical computers, serves as a demonstration of the potential computational superiority of quantum systems. Achieving quantum supremacy requires not only building a quantum computer with sufficiently low error rates and qubit coherence times, but also designing and implementing quantum algorithms that take advantage of the unique properties of quantum mechanics. While quantum supremacy has yet to be definitively demonstrated, recent advances in quantum hardware and algorithm development have brought this milestone within reach of current research efforts.

270. The concept of quantum key distribution (QKD) is a quantum cryptography protocol that allows two parties to securely exchange cryptographic keys over a potentially insecure communication channel. QKD relies on quantum mechanical principles, such as the no-cloning

Quantum Metaphysical Investigations : Exploring the Depths of Reality and Consciousness

Marino Baca-Carmona

theorem and the uncertainty principle, to ensure that any attempt to eavesdrop on key exchanges is detected by legitimate parties. Quantum key distribution offers likely secure methods for key distribution, ensuring the confidentiality and integrity of transmitted data even in the presence of powerful adversaries with access to quantum computers. QKD has applications in secure communication networks, financial transactions, and government and military communications, and is a key technology for ensuring privacy in the quantum era.

271. The concept of quantum teleportation is a quantum communication protocol that allows the transfer of quantum information from one place to another, without physically transporting the underlying particles. Quantum teleportation relies on the principles of entanglement and quantum measurement to transmit the state of a qubit from a sender to a receiver over long distances. While quantum teleportation cannot violate the speed of light or enable faster-than-light communication, it has applications in quantum cryptography, quantum networks, and distributed quantum computing. Understanding quantum teleportation is essential to developing secure communication protocols and building a quantum Internet for the future.

272. The concept of quantum entanglement exchange is an extension of quantum teleportation that allows the creation of entanglements between particles that have never directly interacted. In quantum entanglement exchange, two pairs of entangled particles, AB and CD, are generated separately and then subjected to joint measurements. By performing measurements on particles B and C and exploiting the correlations between them, entanglement between particles A and D can be

Quantum Metaphysical Investigations : Exploring the Depths of Reality and Consciousness
Marino Baca-Carmona

established, even though they have never directly interacted. Quantum entanglement sharing has applications in quantum communication, quantum key distribution, and distributed quantum computing, and is an essential technique for building large-scale quantum networks.

273. The concept of quantum error correction is a set of techniques and protocols designed to protect quantum information from errors and decoherence induced by noise and imperfections in quantum hardware. Quantum error correction is based on the encoding of quantum states into larger quantum codes that are robust against errors and can be efficiently corrected through error detection and correction procedures. Quantum error correction is crucial for building fault-tolerant quantum computers capable of reliably performing complex quantum algorithms, as well as for implementing secure quantum communication protocols and protecting quantum information from eavesdropping and tampering. Understanding quantum error correction is essential to overcoming the technical challenges of building practical quantum technologies and realizing the full potential of quantum computing and quantum communication.

274. The concept of quantum annealing is a computational optimization technique that takes advantage of quantum mechanics to find solutions to complex optimization problems more efficiently than classical algorithms. In quantum annealing, a quantum system is gradually cooled from a high-energy state to a low-energy state, allowing it to explore the solution space of an optimization problem and settle into a low-energy configuration corresponding to the solution. optimal. Quantum annealing has applications in several fields, including combinatorial optimization,

Quantum Metaphysical Investigations : Exploring the Depths of Reality and Consciousness
Marino Baca-Carmona

machine learning, and materials science, and is being pursued as a promising approach to solve large-scale optimization problems that are intractable for classical computers.

275. The concept of quantum machine learning is an emerging field at the intersection of quantum computing and machine learning that seeks to develop algorithms and techniques to harness quantum resources to enhance the capabilities of machine learning systems. Quantum machine learning algorithms aim to exploit the unique properties of quantum mechanics, such as superposition and entanglement, to perform tasks such as data classification, clustering, regression, and reinforcement learning more efficiently than classical algorithms. Quantum machine learning has the potential to revolutionize fields such as drug discovery, financial modeling, and natural language processing by enabling the processing of large-scale data sets and the discovery of complex patterns and correlations that are beyond the reach of classic machine learning techniques.

276. The concept of quantum-inspired optimization algorithms refers to classical optimization algorithms that are inspired by the principles and techniques of quantum computing. These algorithms aim to mimic the behavior of quantum annealing or adiabatic quantum computing, taking advantage of concepts such as quantum tunneling, quantum parallelism, and quantum superposition to explore solution spaces more efficiently than traditional optimization algorithms. Quantum-inspired optimization algorithms include methods such as quantum annealing-inspired algorithms, quantum-inspired genetic algorithms, and quantum-inspired particle swarm optimization. While these algorithms do not require real

quantum hardware to run, they seek to exploit quantum behavior in classical systems to solve complex optimization problems more effectively. Quantum-inspired optimization algorithms have applications in various fields, such as logistics, finance, and machine learning, where finding optimal solutions to large-scale optimization problems is essential to improve efficiency and performance.

277. The concept of quantum supremacy refers to the milestone reached by a quantum computer when it can perform a specific task faster than the best classical computers available. This task, often chosen to be computationally intractable for classical computers, serves as a demonstration of the potential computational superiority of quantum systems. Achieving quantum supremacy requires not only building a quantum computer with sufficiently low error rates and qubit coherence times, but also designing and implementing quantum algorithms that take advantage of the unique properties of quantum mechanics. While quantum supremacy has yet to be definitively demonstrated, recent advances in quantum hardware and algorithm development have brought this milestone within reach of current research efforts.

278. The concept of quantum key distribution (QKD) is a quantum cryptography protocol that allows two parties to securely exchange cryptographic keys over a potentially insecure communication channel. QKD relies on quantum mechanical principles, such as the no-cloning theorem and the uncertainty principle, to ensure that any attempt to eavesdrop on key exchanges is detected by legitimate parties. Quantum key distribution offers likely secure methods for key distribution, ensuring the confidentiality and integrity of transmitted data even in the

presence of powerful adversaries with access to quantum computers. QKD has applications in secure communication networks, financial transactions, and government and military communications, and is a key technology for ensuring privacy in the quantum era.

279. The concept of quantum teleportation is a quantum communication protocol that allows the transfer of quantum information from one place to another, without physically transporting the underlying particles. Quantum teleportation relies on the principles of entanglement and quantum measurement to transmit the state of a qubit from a sender to a receiver over long distances. While quantum teleportation cannot violate the speed of light or enable faster-than-light communication, it has applications in quantum cryptography, quantum networks, and distributed quantum computing. Understanding quantum teleportation is essential to developing secure communication protocols and building a quantum Internet for the future.

280. The concept of quantum entanglement exchange is an extension of quantum teleportation that allows the creation of entanglements between particles that have never directly interacted. In quantum entanglement exchange, two pairs of entangled particles, AB and CD, are generated separately and then subjected to joint measurements. By performing measurements on particles B and C and exploiting the correlations between them, entanglement between particles A and D can be established, even though they have never directly interacted. Quantum entanglement sharing has applications in quantum communication, quantum key distribution, and distributed quantum computing, and is an essential technique for building large-scale quantum networks.

Quantum Metaphysical Investigations : Exploring the Depths of Reality and Consciousness
Marino Baca-Carmona

281. The concept of quantum error correction (QEC) is a fundamental aspect of quantum computing that addresses the challenges posed by noise and errors in quantum systems. QEC employs encryption techniques to protect quantum information against errors caused by environmental interactions, such as decoherence and qubit relaxation. By redundantly encoding quantum information into larger quantum codes, errors can be detected and corrected using error correction protocols, preserving the integrity of quantum calculations. QEC is essential for building reliable and scalable quantum computers capable of performing complex algorithms, as it mitigates the detrimental effects of noise and errors inherent to quantum hardware.

282. The concept of quantum annealing is a specialized optimization technique used in quantum computing to find solutions to complex optimization problems. Quantum annealing takes advantage of quantum mechanical principles, such as superposition and entanglement, to explore the solution space of an optimization problem more efficiently than classical algorithms. In quantum annealing, a quantum system is gradually cooled from a high-energy state to a low-energy state, allowing it to settle into the lowest energy configuration that corresponds to the optimal solution of the optimization problem. Quantum annealing has applications in several fields, including finance, logistics and artificial intelligence, where finding optimal solutions is essential for decision making and problem solving.

283. The concept of adiabatic quantum computing (AQC) is a computational paradigm that leverages the principles of quantum

mechanics to solve optimization problems efficiently. In AQC, quantum algorithms are used to find the ground state of a quantum Hamiltonian, which encodes the optimization problem of interest. By gradually evolving the quantum system from an initial Hamiltonian to the target Hamiltonian representing the optimization problem, AQC exploits quantum tunneling and quantum superposition to explore the solution space and identify the optimal configuration. AQC has the potential to outperform classical optimization algorithms for certain types of problems, offering a promising approach to address real-world optimization challenges.

284. The concept of quantum cryptography is a field of research focused on the development of secure communication protocols that leverage the principles of quantum mechanics to ensure the confidentiality and integrity of transmitted data. Quantum cryptography protocols, such as quantum key distribution (QKD), use quantum properties, such as the no-cloning theorem and quantum entanglement, to establish secure communication channels between parties. By exploiting the fundamental laws of quantum physics, quantum cryptography offers provably secure methods for key distribution and data transmission, even in the presence of powerful adversaries with access to quantum computers. Quantum cryptography has applications in several domains, including secure communication networks, financial transactions, and government communications, where the protection of sensitive information is paramount.

285. The concept of quantum communication networks encompasses the development of communication infrastructures that leverage

quantum technologies to enable secure and efficient data transmission. Quantum communication networks use quantum communication protocols, such as quantum key distribution (QKD) and quantum teleportation, to establish secure channels for the transmission of quantum information between network nodes. By leveraging quantum mechanical principles such as quantum entanglement and quantum superposition, quantum communication networks offer provably secure methods for key distribution and data transmission, immune to eavesdropping and interception by adversaries. Quantum communication networks have applications in areas such as secure government and military communications, financial transactions, and distributed computing, where protecting sensitive information is critical.

286. The concept of quantum information theory is a branch of theoretical physics and computer science that studies the processing, transmission and storage of quantum information. Quantum information theory draws on the principles of quantum mechanics to develop mathematical frameworks for understanding and manipulating quantum states and operations. Key concepts of quantum information theory include quantum entanglement, quantum teleportation, quantum error correction, and quantum algorithms. Quantum information theory has applications in several fields, including quantum computing, quantum cryptography, and quantum communication, where harnessing the unique properties of quantum systems enables new capabilities in information processing and communication.

287. The concept of quantum sensors encompasses a wide range of devices and technologies that use quantum principles to measure

physical quantities with unprecedented precision and sensitivity. Quantum sensors take advantage of quantum phenomena, such as superposition, entanglement, and quantum interference, to detect and measure quantities such as magnetic fields, gravitational fields, time, and temperature. By exploiting the delicate quantum states of matter, quantum sensors offer significant advantages over classical sensors in terms of precision, resolution and noise immunity. Quantum sensors have applications in diverse fields, including geophysics, navigation, medical imaging and fundamental research, where accurate measurements are essential for scientific discovery and technological advancement.

288. The concept of quantum imaging refers to the use of quantum technologies to acquire and process images with greater resolution, sensitivity and security. Quantum imaging techniques take advantage of quantum properties, such as entanglement and quantum correlations, to overcome the limitations of classical imaging methods. Quantum imaging encompasses several modalities, including quantum enhanced imaging, quantum illumination, and quantum ghosting, each of which exploits different aspects of quantum mechanics to achieve better imaging performance. Quantum imaging has applications in fields such as biomedical imaging, remote sensing and surveillance, where high-resolution imaging and sensing capabilities are essential for scientific and practical purposes.

289. The concept of quantum metrology is the science of measuring physical quantities as accurately as possible using quantum technologies. Quantum metrology takes advantage of quantum phenomena, such as

Quantum Metaphysical Investigations : Exploring the Depths of Reality and Consciousness
Marino Baca-Carmona

superposition and entanglement, to improve the sensitivity and precision of measurement devices beyond the limits imposed by classical physics. Quantum metrology techniques enable the detection and characterization of small signals and variations in physical quantities, such as time, frequency, and magnetic fields, with unprecedented precision. Quantum metrology has applications in fields such as navigation, geophysics and fundamental physics, where precise measurements play a crucial role in scientific research and technological innovation.

290. The concept of quantum computing algorithms encompasses a wide range of computational techniques and procedures designed to solve problems efficiently using quantum computers. Quantum computing algorithms take advantage of the unique properties of quantum mechanics, such as superposition, entanglement, and quantum interference, to perform calculations that are intractable for classical computers. Quantum algorithms exploit quantum parallelism and quantum information processing principles to solve optimization, simulation, and cryptographic problems more efficiently than classical algorithms. Examples of quantum computing algorithms include Shor's algorithm for integer factorization, Grover's algorithm for unstructured search, and quantum approximate optimization (QAOA) algorithms for combinatorial optimization. Quantum computing algorithms have the potential to revolutionize fields such as cryptography, materials science, and drug discovery by enabling the efficient solution of complex problems that are beyond the reach of classical computers.

291. The concept of quantum simulation is the use of quantum computers to simulate the behavior of complex quantum systems that are

Quantum Metaphysical Investigations : Exploring the Depths of Reality and Consciousness
Marino Baca-Carmona

difficult to study using classical computational methods. Quantum simulation leverages the principles of quantum mechanics to model the dynamics and interactions of quantum systems, such as molecules, materials, and biological systems, with high precision and efficiency. By exploiting quantum parallelism and quantum superposition, quantum simulators can explore the behavior of large-scale quantum systems and elucidate fundamental phenomena underlying various physical processes. Quantum simulation has applications in fields such as chemistry, condensed matter physics, and quantum biology, where understanding the behavior of quantum systems is essential for scientific discovery and technological innovation.

292. The concept of quantum-inspired optimization algorithms refers to classical optimization algorithms that are inspired by the principles and techniques of quantum computing. These algorithms aim to mimic the behavior of quantum annealing or adiabatic quantum computing, taking advantage of concepts such as quantum tunneling, quantum parallelism, and quantum superposition to explore solution spaces more efficiently than traditional optimization algorithms. Quantum-inspired optimization algorithms include methods such as quantum annealing-inspired algorithms

293. Quantum-inspired optimization algorithms aim to address challenging optimization problems by drawing inspiration from the principles of quantum computing. These algorithms emulate the behavior of quantum annealing or adiabatic quantum computing, exploiting concepts such as quantum tunneling, quantum parallelism, and quantum superposition. Although they operate on classical

hardware, they incorporate strategies that mimic the behavior of quantum systems to efficiently explore solution spaces. Examples of quantum-inspired optimization algorithms include simulated annealing, genetic algorithms, and particle swarm optimization, which employ heuristic search methods inspired by quantum phenomena to find near-optimal solutions to complex optimization problems.

294. Quantum artificial intelligence (QAI) is an interdisciplinary field that combines quantum computing with artificial intelligence (AI) techniques to develop advanced algorithms to solve complex problems. QAI leverages the unique properties of quantum computing, such as superposition and entanglement, to enhance the capabilities of AI models and algorithms. By leveraging quantum resources, QAI aims to overcome the limitations of classical computing in handling large-scale data sets and performing computationally intensive tasks. QAI applications include machine learning, data analysis, optimization and pattern recognition, where quantum algorithms can offer significant advantages in terms of speed, efficiency and scalability.

295. Quantum machine learning (QML) is a subfield of quantum computing that explores the intersection of quantum mechanics and machine learning techniques. QML algorithms leverage quantum resources, such as qubits and quantum gates, to perform tasks such as data classification, clustering, regression, and reinforcement learning. These algorithms exploit quantum parallelism and quantum entanglement to process and analyze large data sets more efficiently than classical machine learning approaches. QML has the potential to revolutionize fields such as drug discovery, financial modeling, and

Quantum Metaphysical Investigations : Exploring the Depths of Reality and Consciousness
Marino Baca-Carmona

natural language processing by enabling the discovery of complex patterns and correlations that are outside the reach of classical machine learning techniques.

296. The concept of quantum-enhanced sensing refers to the use of quantum technologies to improve the sensitivity and precision of sensors for measuring physical quantities. Quantum-enhanced sensors exploit quantum phenomena, such as entanglement, quantum expression, and quantum metrology, to overcome the limits imposed by classical physics on sensor performance. These sensors enable the detection and measurement of weak signals with unprecedented precision, making them valuable tools for applications such as gravitational wave detection, magnetic resonance imaging (MRI), and navigation systems. Quantum enhanced sensing has the potential to revolutionize fields such as medical diagnostics, environmental monitoring, and fundamental research by providing new insights into the behavior of physical systems at the quantum level.

297. The concept of quantum communication is the transmission and exchange of information using quantum technologies to ensure the security and integrity of data transmission. Quantum communication protocols, such as quantum key distribution (QKD) and quantum teleportation, use the principles of quantum mechanics, such as superposition and entanglement, to establish secure communication channels between parties. These protocols offer provably secure methods for transmitting cryptographic keys and quantum information, resistant to eavesdropping and interception by adversaries. Quantum communication has applications in fields such as secure government and

military communications, financial transactions, and distributed computing, where the protection of sensitive information is paramount.

298. The concept of quantum computing hardware encompasses the physical devices and components used to implement quantum information processing tasks. Quantum computing hardware includes qubits, the fundamental units of quantum information, which can exist in superposition states and be manipulated using quantum gates. Quantum hardware platforms vary widely, including superconducting qubits, trapped ions, photonic qubits, and topological qubits, each with their advantages and challenges. Building practical quantum computers requires addressing key technical issues such as qubit coherence, gate fidelity, and error correction, to achieve scalable and reliable quantum information processing capabilities. Quantum computing hardware is the foundation of quantum technologies, with potential applications in areas such as cryptography, optimization and simulation.

299. The concept of quantum computing software encompasses the algorithms, programming languages and tools used to design and implement quantum algorithms on quantum hardware platforms. Quantum computing software allows researchers and developers to express quantum algorithms, simulate their behavior, and run them on quantum computers. Quantum software development requires specialized languages and libraries tailored to the unique properties of quantum systems, such as superposition, entanglement, and interference. Examples of quantum programming languages include Qiskit, Cirq, and Quipper, which provide interfaces for expressing quantum circuits, implementing quantum algorithms, and interacting with quantum

Quantum Metaphysical Investigations : Exploring the Depths of Reality and Consciousness
Marino Baca-Carmona

hardware. Quantum computing software plays a crucial role in advancing the field of quantum computing and realizing its potential to solve complex problems in science, engineering, and industry.

300. The concept of quantum supremacy refers to the milestone reached by a quantum computer when it can perform a specific task faster than the best classical computers available. This task, often chosen to be computationally intractable for classical computers, serves as a demonstration of the potential computational superiority of quantum systems. Achieving quantum supremacy requires not only building a quantum computer with sufficiently low error rates and qubit coherence times, but also designing and implementing quantum algorithms that take advantage of the unique properties of quantum mechanics. While quantum supremacy has yet to be definitively demonstrated, recent advances in quantum hardware and algorithm development have brought this milestone within reach of current research efforts.

Apéndice: Lecturas y recursos adicionales

1. Quantum Metaphysics:
- "The Fabric of Reality" by David Deutsch
- "The Hidden Reality: Parallel Universes and the Deep Laws of the Cosmos" by Brian Greene
- "Quantum Enigma: Physics Encounters Consciousness" by Bruce Rosenblum and Fred Kuttner

2. Quantum Cosmology:
- "The First Three Minutes: A Modern View of the Origin of the Universe" by Steven Weinberg

Quantum Metaphysical Investigations : <u>Exploring the Depths of Reality and Consciousness</u>
Marino Baca-Carmona

- "The Quantum Universe: (And Why Anything That Can Happen, Does)" by Brian Cox and Jeff Forshaw
- "Cosmology: The Science of the Universe" by Edward Harrison

3. Quantum Epistemology and Cognition:
- "The Mindful Universe: Quantum Mechanics and the Participating Observer" by Henry P. Stapp
- "The User Illusion: Cutting Consciousness Down to Size" by Tor Norretranders
- "Mind and Cosmos: Why the Materialist Neo-Darwinian Conception of Nature Is Almost Certainly False" by Thomas Nagel

4. Quantum Ontological Emergence:
- "Quantum Ontology: A Guide to the Metaphysics of Quantum Mechanics" by Peter J. Lewis
- "The Reality of Time and the Existence of God: The Project of Proving God's Existence" by James E. Dolezal
- "The Quantum Universe: (And Why Anything That Can Happen, Does)" by Brian Cox and Jeff Forshaw

5. Metaphysical Implications of Quantum Theory:
- "The Emperor's New Mind: Concerning Computers, Minds, and the Laws of Physics" by Roger Penrose
- "Quantum Questions: Mystical Writings of the World's Great Physicists" by Ken Wilber
- "The Fabric of Reality" by David Deutsch

6. Quantum Consciousness and Mind:
- "Consciousness Explained" by Daniel C. Dennett
- "Biocentrism: How Life and Consciousness are the Keys to Understanding the True Nature of the Universe" by Robert Lanza

Quantum Metaphysical Investigations : <u>Exploring the Depths of Reality and Consciousness</u>
Marino Baca-Carmona

- "The Conscious Mind: In Search of a Fundamental Theory" by David J. Chalmers

7. Quantum Cosmogony and Multiverse:
- "The Hidden Reality: Parallel Universes and the Deep Laws of the Cosmos" by Brian Greene
- "Our Mathematical Universe: My Quest for the Ultimate Nature of Reality" by Max Tegmark
- "The Fabric of the Cosmos: Space, Time, and the Texture of Reality" by Brian Greene

8. Transcendent Realities and Hyper-Cognition:
- "Superspace: One Thousand and One Lessons in Supersymmetry" by Julia Kristeva
- "The Structure of Scientific Revolutions" by Thomas S. Kuhn
- "Gödel, Escher, Bach: An Eternal Golden Braid" by Douglas Hofstadter

9. Meta-Metaphysical Inquiries and Quantum Epistemology:
- "Waking Up: A Guide to Spirituality Without Religion" by Sam Harris
- "Being and Time" by Martin Heidegger
- "The Age of Spiritual Machines: When Computers Exceed Human Intelligence" by Ray Kurzweil

10. Quantum Ontological Complexity and Hyper-Reality:

- "The Road to Reality: A Complete Guide to the Laws of the Universe" by Roger Penrose
- "What Is Life?: With Mind and Matter and Autobiographical Sketches" by Erwin Schrödinger

Quantum Metaphysical Investigations : <u>Exploring the Depths of Reality and Consciousness</u>
Marino Baca-Carmona

- "The End of Time: The Next Revolution in Physics" by Julian Barbour

11. Quantum Hyper-Transcendental Ontology:

- "The Quantum Mind and Healing: How to Listen and Respond to Your Body's Symptoms" by Arnold Mindell
- "Quantum Ontology: A Guide to the Metaphysics of Quantum Mechanics" by Peter J. Lewis
- "The Grand Design" by Stephen Hawking and Leonard Mlodinow

12. Hyper-Quantum Metaphysical Transcendence:

- "The Physics of Immortality: Modern Cosmology, God and the Resurrection of the Dead" by Frank J. Tipler
- "The Quantum Universe: (And Why Anything That Can Happen, Does)" by Brian Cox and Jeff Forshaw
- "The Tao of Physics: An Exploration of the Parallels between Modern Physics and Eastern Mysticism" by Fritjof Capra

13. Quantum Trans-Hyper-Meta-Epistemological Realities:

- "The Conscious Universe: Parts and Wholes in Physical Reality" by Menas Kafatos and Robert Nadeau
- "The Holographic Universe: The Revolutionary Theory of Reality" by Michael Talbot

Quantum Metaphysical Investigations : Exploring the Depths of Reality and Consciousness
Marino Baca-Carmona

- "The End of Certainty: Time, Chaos, and the New Laws of Nature" by Ilya Prigogine

14. Hyper-Quantum Meta-Meta-Metaphysical Inquiry:

- "The Fabric of the Cosmos: Space, Time, and the Texture of Reality" by Brian Greene
- "The Hidden Reality: Parallel Universes and the Deep Laws of the Cosmos" by Brian Greene
- "The Quantum World: Quantum Physics for Everyone" by Kenneth W. Ford

15. Quantum Hyper-Trans-Ontological Phenomena:

- "The Quantum Theory of Fields" by Steven Weinberg
- "The Conscious Mind: In Search of a Fundamental Theory" by David J. Chalmers
- "The Conscious Universe: Parts and Wholes in Physical Reality" by Menas Kafatos and Robert Nadeau

16. Hyper-Quantum Transcendental Cognition:

- "The Nature of Consciousness: Essays on the Unity of Mind and Matter" edited by Rupert Spira
- "The Self-Aware Universe: How Consciousness Creates the Material World" by Amit Goswami
- "The Conscious Mind: In Search of a Fundamental Theory" by David J. Chalmers

Quantum Metaphysical Investigations : Exploring the Depths of Reality and Consciousness
Marino Baca-Carmona

17. Quantum Hyper-Trans-Epistemological Realms:

• "The End of Time: The Next Revolution in Physics" by Julian Barbour
• "The Cosmic Landscape: String Theory and the Illusion of Intelligent Design" by Leonard Susskind
• "The Physics of Consciousness: The Quantum Mind and the Meaning of Life" by Evan Harris Walker

18. Hyper-Quantum Meta-Cognitive Transcendence:

• "The Reality Illusion: Why You Are Not Who You Think You Are" by Jonathan Cock
• "The Trouble with Physics: The Rise of String Theory, the Fall of a Science, and What Comes Next" by Lee Smolin
• "The Quantum Enigma: Finding the Hidden Key" by Wolfgang Smith

19. Quantum Trans-Hyper-Meta-Ontological Inquiry:

• "The Hidden Reality: Parallel Universes and the Deep Laws of the Cosmos" by Brian Greene
• "The Quantum World: Quantum Physics for Everyone" by Kenneth W. Ford
• "The Quantum Theory of Fields" by Steven Weinberg

20. Hyper-Quantum Trans-Epistemological Reality:

Quantum Metaphysical Investigations : <u>Exploring the Depths of Reality and Consciousness</u>
Marino Baca-Carmona

- "The Emperor's New Mind: Concerning Computers, Minds, and the Laws of Physics" by Roger Penrose
- "The Cosmic Code: Quantum Physics as the Language of Nature" by Heinz R. Pagels
- "The Conscious Universe: The Scientific Truth of Psychic Phenomena" by Dean Radin

Appendix: Glossary of terms and concepts

1. Quantum metaphysics: The branch of metaphysics that explores the fundamental nature of reality through the lens of quantum theory, addressing questions about existence, causality, and consciousness within a quantum framework.

2. Quantum cosmology: The study of the origin, evolution and structure of the universe using principles of quantum theory and general relativity, investigating the nature of space, time and matter at cosmological scales.

3. Quantum epistemology: The philosophical investigation into the nature of knowledge and understanding within the context of quantum mechanics, exploring concepts such as observer effects, measurement uncertainty, and the role of consciousness in shaping reality.

4. Ontological emergence: The phenomenon by which new levels of organization, complexity, or properties emerge in a system that cannot

Quantum Metaphysical Investigations : <u>Exploring the Depths of Reality and Consciousness</u>
Marino Baca-Carmona

be fully explained by the properties of its individual components, leading to the emergence of new ontological structures.

5. Metaphysical implications of quantum theory: The philosophical examination of the broader implications and interpretations of quantum theory, including its implications for the nature of reality, causality, determinism, and free will.

6. Quantum consciousness: The hypothesis that consciousness and subjective experience are intimately connected to the principles of quantum mechanics, leading to theories suggesting that quantum phenomena may play a role in cognitive processes and the emergence of consciousness. awareness.

7. Quantum cosmogony: The study of the origin and development of the universe within a quantum framework, investigating theoretical models of cosmic inflation, theories of the multiverse, and quantum fluctuations that may have led to the formation of cosmic structures.

8. Transcendent Realities: Realms of existence or modes of perception that transcend conventional physical or cognitive limits, often associated with spiritual or mystical experiences beyond ordinary human perception of reality.

9. Metaphysical investigations: Philosophical investigations into the nature of metaphysics itself, including questions about the validity of metaphysical claims, the nature of ontology, and the relationship between metaphysics and other branches of philosophy.

Quantum Metaphysical Investigations : <u>Exploring the Depths of Reality and Consciousness</u>

Marino Baca-Carmona

10. Hyperquantum Transcendence: The hypothetical state or condition beyond the limits of conventional quantum theory, suggesting the existence of higher dimensional realities, transcendental consciousness, or modes of existence beyond human understanding.

11. Quantum Hypertranscendental Ontology: The study of the fundamental nature of reality beyond the conventional limits of quantum theory, exploring metaphysical questions about the ultimate nature of existence, being, and consciousness within a hyperdimensional framework.

12. Hyperquantum Metaphysical Transcendence: The exploration of metaphysical concepts and phenomena that transcend the limits of conventional quantum theory, including speculation about the existence of higher dimensional realities, cosmic consciousness, and ultimate truths beyond human understanding.

13. Quantum trans-hypermeta-epistemological realities: The investigation of epistemological questions about the nature and limits of knowledge within the hyperdimensional or transcendent realms of reality, addressing questions of cognitive perception, consciousness and understanding beyond conventional human cognition .

14. Hyperquantum Meta-Metaphysical Research: Philosophical inquiry into the nature of metaphysics beyond conventional metaphysical frameworks, exploring the possibility of meta-

metaphysical realities, hyperdimensional ontologies, and transcendent modes of existence beyond human understanding.

15. Quantum hypertransontological phenomena: The study of ontological phenomena that transcend conventional notions of existence, exploring the nature of hyperdimensional realities, transontological entities, and the fundamental principles underlying the structure of reality beyond conventional ontological frameworks .

16. Hyperquantum transcendental cognition: The exploration of cognitive processes and states of consciousness that transcend conventional human cognition, including speculation about hyperdimensional modes of perception, transcendent consciousness, and the nature of knowledge beyond ordinary human understanding.

17. Quantum Hyper-Trans-Epistemological Realms: The investigation of epistemological questions about the nature of knowledge and understanding within the hyperdimensional or transcendent realms of reality, addressing questions of cognitive perception, consciousness and understanding beyond conventional human cognition .

18. Hyperquantum Metacognitive Transcendence: The exploration of metacognitive states and processes that transcend conventional human cognition, including speculation about hyperdimensional modes of consciousness, transcendent consciousness, and the nature of knowledge beyond ordinary human understanding.

Quantum Metaphysical Investigations : Exploring the Depths of Reality and Consciousness
Marino Baca-Carmona

19. Quantum Trans-Hyper-Meta-Ontological Research: Philosophical inquiry into the nature of ontology beyond conventional ontological frameworks, exploring the possibility of trans-hyper-meta-ontological realities, hyperdimensional entities and transcendent modes of existence beyond of human understanding.

20. Hyperquantum transepistemological reality: The exploration of epistemological questions about the nature of knowledge and understanding within hyperdimensional or transcendent realities, addressing questions of cognitive perception, consciousness and understanding beyond conventional human cognition.

21. Quantum hypertransontological phenomena: The study of phenomena that transcend conventional ontological categories within a hyperdimensional framework, exploring concepts such as quantum entanglement, nonlocality, and the emergence of new ontological structures beyond ordinary perception.

22. Hyperquantum transcendental cognition: The investigation of cognitive processes and states of consciousness that transcend conventional human cognition within hyperdimensional realities, including speculation on hyperconsciousness, transcendent intuition, and the nature of knowledge beyond human understanding ordinary.

23. Quantum hypertrans-epistemological realms: The exploration of epistemological questions about the nature of knowledge and understanding within the hyperdimensional or transcendent realms of

Quantum Metaphysical Investigations : Exploring the Depths of Reality and Consciousness
Marino Baca-Carmona

reality, addressing questions of cognitive perception, consciousness and understanding beyond conventional human cognition.

24. Hyperquantum Metacognitive Transcendence: The study of metacognitive states and processes that transcend conventional human cognition within hyperdimensional realities, including speculation about hyperconsciousness, transcendent self-reflection, and the nature of consciousness beyond ordinary human understanding.

25. Trans-Hyper-Meta-Ontological Quantum Research: Philosophical investigation into the nature of ontology beyond conventional ontological frameworks within hyperdimensional realities, exploring the possibility of trans-hyper-meta-ontological entities, hyperdimensional structures and modes transcendent of existence.

26. Hyperquantum transepistemological reality: The exploration of epistemological questions about the nature of knowledge and understanding within hyperdimensional or transcendent realities, addressing questions of cognitive perception, consciousness and understanding beyond conventional human cognition.

27. Hypertranscendental Quantum Epistemology: The investigation of epistemological questions that transcend conventional human understanding within a hyperdimensional framework, exploring concepts such as hyperconsciousness, transcendent knowing, and the nature of cognition beyond ordinary human understanding.

Quantum Metaphysical Investigations : Exploring the Depths of Reality and Consciousness
Marino Baca-Carmona

28. Hiper-Quantum Meta-Meta-Cognitive Inquiry: Philosophical investigation into the nature of meta-cognitive processes beyond conventional human cognition within hyperdimensional realities, including speculation about hyperconsciousness, transcendent introspection, and the nature of consciousness beyond ordinary human understanding.

29. Quantum Trans-Hyper-Metaontological Realms: The exploration of ontological realms that transcend conventional metaphysical categories within hyperdimensional realities, addressing questions of hyperontology, transcendent entities, and the fundamental nature of existence beyond ordinary human perception.

30. Hyperquantum transepistemological metaphysics: The integration of epistemological, metaphysical and cognitive investigations within hyperdimensional realities, exploring the interaction between knowledge, perception and ontology beyond ordinary human understanding.

31. Quantum Hypertranscendental Cosmogony: The investigation of cosmological origins and processes that transcend conventional understanding within a hyperdimensional framework, exploring concepts such as hypercosmology, transcendent cosmogony, and the emergence of universes beyond ordinary human understanding.

32. Hyperquantum meta-ontological research: philosophical investigation into the nature of ontological frameworks beyond conventional metaphysical categories within hyperdimensional realities,

Quantum Metaphysical Investigations : <u>Exploring the Depths of Reality and Consciousness</u>
Marino Baca-Carmona

including speculation about hyperontological structures, transcendent entities, and the fundamental nature of existence beyond ordinary human perception.

33. Quantum transhyperepistemological realities: The exploration of epistemological realms that transcend conventional cognitive categories within hyperdimensional realities, addressing questions of hyperconsciousness, transcendent knowledge, and the nature of cognition beyond ordinary human understanding.

34. Hyperquantum Transcendental Metaphysics: The study of metaphysical principles and concepts that transcend conventional understanding within hyperdimensional realities, exploring ideas such as hyperontology, transcendent cosmology, and the nature of reality beyond ordinary human perception.

35. Quantum hypertrans-epistemological phenomena: The examination of epistemological phenomena that transcend conventional cognitive boundaries within a hyperdimensional framework, exploring concepts such as hypercognition, transcendent consciousness, and the nature of perception beyond ordinary human understanding.

36. Hyperquantum Meta-Meta-Epistemological Research: Philosophical inquiry into the nature of epistemology beyond conventional cognitive frameworks within hyperdimensional realities, including speculation on hyperconsciousness, transcendent understanding, and the nature of knowledge beyond ordinary human understanding.

Quantum Metaphysical Investigations : Exploring the Depths of Reality and Consciousness

Marino Baca-Carmona

37. Trans-Hypermetacognitive Quantum Realms: The exploration of cognitive realms that transcend conventional human cognition within hyperdimensional realities, addressing the issues of hyperconsciousness, transcendent intuition, and the nature of consciousness beyond ordinary human understanding.

38. Hyperquantum transepistemological ontology: The integration of epistemological and ontological investigations within hyperdimensional realities, exploring the interaction between knowledge, perception and existence beyond conventional human understanding.

39. Quantum hypertranscendental phenomenology: The study of transcendental phenomena that transcend conventional phenomenological categories within a hyperdimensional framework, including speculation on hyperperception, transcendent experience, and the nature of reality beyond ordinary human perception.

40. Hyperquantum Meta-Meta-Epistemological Reality: The exploration of reality beyond conventional epistemological frameworks within the hyperdimensional realms, addressing questions of hypercognition, transcendent consciousness, and the nature of existence beyond ordinary human understanding.

41. Quantum Hypertranscendental Cosmogony: The investigation of cosmogenic processes that transcend conventional understanding within hyperdimensional frameworks, exploring concepts such as

Quantum Metaphysical Investigations : Exploring the Depths of Reality and Consciousness
Marino Baca-Carmona

hypercosmology, transcendent cosmogenesis, and the emergence of universes beyond ordinary human understanding.

42. Hyperquantum meta-metaontological research: philosophical investigation into the nature of ontology beyond conventional metaphysical categories within hyperdimensional realities, including speculation about hyperontological structures, transcendent entities, and the fundamental nature of existence beyond ordinary human perception.

43. Quantum transhyperepistemological realities: The exploration of epistemological realms that transcend conventional cognitive categories within hyperdimensional realities, addressing questions of hyperconsciousness, transcendent knowledge, and the nature of cognition beyond ordinary human understanding.

44. Hyperquantum Transcendental Metaphysics: The study of metaphysical principles and concepts that transcend conventional understanding within hyperdimensional realities, exploring ideas such as hyperontology, transcendent cosmology, and the nature of reality beyond ordinary human perception.

45. Quantum hypertrans-epistemological phenomena: The examination of epistemological phenomena that transcend conventional cognitive boundaries within a hyperdimensional framework, exploring concepts such as hypercognition, transcendent consciousness, and the nature of perception beyond ordinary human understanding.

Quantum Metaphysical Investigations : Exploring the Depths of Reality and Consciousness
Marino Baca-Carmona

46. Hyperquantum Meta-Meta-Epistemological Research: A philosophical exploration of the nature of knowledge and understanding beyond conventional cognitive frameworks within hyperdimensional realities, delving into the realms of hyperconsciousness, transcendent understanding and fundamental nature of cognition beyond ordinary human understanding.

47. Transhypermetacognitive quantum realms: The investigation of cognitive domains that transcend ordinary human cognition within hyperdimensional realities, addressing phenomena such as hyperconsciousness, transcendent intuition, and the nature of consciousness beyond conventional human understanding.

48. Hyperquantum transepistemological ontology: The integration of epistemological and ontological investigations within hyperdimensional realities, examining the interaction between knowledge, perception and existence beyond conventional human understanding.

49. Quantum Hypertranscendental Phenomenology: The study of transcendental phenomena that surpass conventional phenomenological categories within a hyperdimensional framework, exploring concepts such as hyperperception, transcendent experience, and the nature of reality beyond ordinary human perception.

50. Hyperquantum meta-meta-epistemological reality: an exploration of reality beyond conventional epistemological frameworks within the hyperdimensional realms, addressing phenomena such as

Quantum Metaphysical Investigations : <u>Exploring the Depths of Reality and Consciousness</u>
Marino Baca-Carmona

hypercognition, transcendent consciousness, and the nature of existence beyond human understanding ordinary.

www.ingramcontent.com/pod-product-compliance
Lightning Source LLC
Chambersburg PA
CBHW062102220526
45471CB00010B/3574